PHYTOCHEMICAL AND PHARMACOLOGICAL INVESTIGATION OF THE FAMILY RUTACEAE

Innovations in Plant Science for Better Health: From Soil to Fork

PHYTOCHEMICAL AND PHARMACOLOGICAL INVESTIGATION OF THE FAMILY RUTACEAE

Edited by

Abdur Rauf, PhD

Hafiz Ansar Rasul Suleria, PhD

Syed Muhammad Mukarram Shah, PhD

APPLE ACADEMIC PRESS

First edition published 2024

Apple Academic Press Inc.
1265 Goldenrod Circle, NE,
Palm Bay, FL 32905 USA

760 Laurentian Drive, Unit 19,
Burlington, ON L7N 0A4, CANADA

CRC Press
2385 NW Executive Center Drive,
Suite 320, Boca Raton FL 33431

4 Park Square, Milton Park,
Abingdon, Oxon, OX14 4RN UK

Library and Archives Canada Cataloguing in Publication

Title: Phytochemical and pharmacological investigation of the family rutaceae / edited by Abdur Rauf, PhD, Hafiz Ansar Rasul Suleria, PhD, Syed Muhammad Mukarram Shah, PhD.
Names: Rauf, Abdur (Lecturer in chemistry), editor. | Suleria, Hafiz, editor. | Shah, Syed Muhammad Mukarram, editor.
Series: Innovations in plant science for better health.
Description: First edition. | Series statement: Innovations in plant science for better health : from soil to fork | Includes bibliographical references and index.
Identifiers: Canadiana (print) 20230556159 | Canadiana (ebook) 20230556205 | ISBN 9781774913710 (hardcover) | ISBN 9781774913727 (softcover) | ISBN 9781003401469 (ebook)
Subjects: LCSH: Rutaceae. | LCSH: Rutaceae—Therapeutic use. | LCSH: Phytochemicals. | LCSH: Medicinal plants. | LCSH: Pharmacognosy.
Classification: LCC QK495.R98 P49 2024 | DDC 583/.75—dc23

Library of Congress Cataloging-in-Publication Data

CIP data on file with US Library of Congress

ISBN: 978-1-77491-371-0 (hbk)
ISBN: 978-1-77491-372-7 (pbk)
ISBN: 978-1-00340-146-9 (ebk)

Books in Innovations in Plant Science for Better Health: From Soil to Fork series

Book Series: *Innovations in Plant Science for Better Health: From Soil to Fork*
Editor-in-Chief: Hafiz Ansar Rasul Suleria, PhD

- **Assessment of Medicinal Plants for Human Health: Phytochemistry, Disease Management, and Novel Applications**
  Editors: Megh R. Goyal, PhD, and Durgesh Nandini Chauhan, MPharm
- **Bioactive Compounds of Medicinal Plants: Properties and Potential for Human Health**
  Editors: Megh R. Goyal, PhD, and Ademola O. Ayeleso
- **Bioactive Compounds from Plant Origin: Extraction, Applications, and Potential Health Claims**
  Editors: Hafiz Ansar Rasul Suleria, PhD, and Colin Barrow, PhD
- **Bioactive Compounds from Multifarious Natural Foods for Human Health**
  Editors: Hafiz Ansar Rasul Suleria, PhD, Megh R. Goyal, PhD, and Huma Bader Ul Ain
- **Cereals and Cereal-Based Foods**
  Editors: Megh R. Goyal, PhD, Kamaljit Kaur, PhD, Jaspreet Kaur, PhD
- **Curative and Preventive Properties of Medicinal Plants: Research on Disease Management and Animal Model Studies**
  Editors: Megh R. Goyal, PhD, Junaid Ahmad Malik, PhD, and Ademola Olabode Ayeleso, PhD
- **Health Benefits of Secondary Phytocompounds from Plant and Marine Sources**
  Editors: Hafiz Ansar Rasul Suleria, PhD, and Megh Goyal, PhD
- **Herbs, Spices, and Medicinal Plants for Human Gastrointestinal Disorders**
  Editors: Megh R. Goyal, PhD, Preeti Birwal, PhD, and Durgesh Nandini Chauhan, MPharm
- **Himalayan Medicinal Plants for the Treatment of Depression: A Source of Rich Antidepressant Agents**
  Editors: Abdur Rauf, PhD, Prabhakar Semwal, PhD,and Hafiz Ansar Rasul Suleria, PhD
- **Human Health Benefits of Plant Bioactive Compounds: Potentials and Prospects**
  Editors: Megh R. Goyal, PhD, and Hafiz Ansar Rasul Suleria, PhD

- **Phytochemical and Pharmacological Investigation of the Family Rutaceae**
  Editors: Abdur Rauf, PhD, Hafiz Ansar Rasul Suleria, PhD, and
  Syed Muhammad Mukarram Shah, PhD
- **Phytochemicals and Medicinal Plants in Food Design: Strategies and Technologies for Improved Healthcare**
  Editors: Megh R. Goyal, PhD, Preeti Birwal, PhD, and Santosh K. Mishra, PhD
- **Phytochemicals from Medicinal Plants: Scope, Applications, and Potential Health Claims**
  Editors: Hafiz Ansar Rasul Suleria, PhD, Megh R. Goyal, PhD, and
  Masood Sadiq Butt, PhD
- **Plant- and Marine-Based Phytochemicals for Human Health: Attributes, Potential, and Use**
  Editors: Megh R. Goyal, PhD, and Durgesh Nandini Chauhan, MPharm
- **Plant-Based Bioactive Compounds and Food Ingredients: Encapsulation, Functional, and Safety Aspects**
  Editors: Junaid Ahmad Malik, PhD, Megh R. Goyal, PhD, Preeti Birwal, PhD, and
  Ritesh B. Watharkar, PhD
- **Plant-Based Functional Foods and Phytochemicals: From Traditional Knowledge to Present Innovation**
  Editors: Megh R. Goyal, PhD, Arijit Nath, PhD, and Hafiz Ansar Rasul Suleria, PhD
- **Plant Secondary Metabolites for Human Health: Extraction of Bioactive Compounds**
  Editors: Megh R. Goyal, PhD, P. P. Joy, PhD, and Hafiz Ansar Rasul Suleria, PhD
- **The Functional Foods: Nutrient and Health Benefits**
  Editors: Megh R. Goyal, PhD, Junaid Ahmad Malik, PhD, and Anu Kumari
- **The Role of Phytoconstitutents in Healthcare: Biocompounds in Medicinal Plants**
  Editors: Megh R. Goyal, PhD, Hafiz Ansar Rasul Suleria, PhD,
  and Ramasamy Harikrishnan, PhD
- **The Therapeutic Properties of Medicinal Plants: Health-Rejuvenating Bioactive Compounds of Native Flora**
  Editors: Megh R. Goyal, PhD, PE, Hafiz Ansar Rasul Suleria, PhD,
  Ademola Olabode Ayeleso, PhD, T. Jesse Joel, and Sujogya Kumar Panda

About Series Editor-in-Chief

Hafiz Suleria, PhD
McKenzie Fellow, Department of Agriculture and Food Systems, The University of Melbourne, Victoria, Australia

Hafiz Suleria, PhD is currently working as McKenzie Fellow, Department of Agriculture and Food Systems, The University of Melbourne, Victoria, Australia.

Dr. Suleria has been awarded an International Postgraduate Research Scholarship (IPRS) and Australian Postgraduate Award (APA) for his PhD research at UQ School of Medicine, the Translational Research Institute (TRI) in collaboration with Commonwealth and Scientific and Industrial Research Organization (CSIRO, Australia).

Before joining the UQ School of Medicine, he worked as a lecturer in the Department of Food Sciences, Government College University Faisalabad, Pakistan. He also worked as research associate in PAK-US Joint Project funded by Higher Education Commission, Pakistan and Department of State, USA in collaboration with the University of Massachusetts, USA and National Institute of Food Science and Technology, University of Agriculture, Faisalabad, Pakistan.

He has a major research focus on food nutrition, particularly in screening of bioactive molecules—isolation, purification, and characterization using various cutting-edge techniques from different plants, marine, and animal sources—in vitro, in vivo bioactivities, cell culture, and animal modeling. He has quite a reasonable work on functional foods and nutraceutical foods and functions, and alternative medicine.

Dr. Suleria has published more than 50 peer-reviewed scientific papers in different reputed/impacted journals. He is also in collaboration with more than 10 universities where he is working as a co-supervisor/special member for PhD and postgraduate students and is also involved in joint publications, projects, and grants.

About the Editors

Abdur Rauf, PhD
Assistant Professor, Department of Chemistry, University of Swabi, Swabi, Anbar, Khyber Pakhtunkhwa, Pakistan

Abdur Rauf, PhD, is at present, an Assistant Professor in the Department of Chemistry, University of Swabi, Swabi, Anbar, Khyber Pakhtunkhwa, Pakistan.

Dr. Abdur Rauf completed his PhD at the Institute of Chemical Sciences, University of Peshawar, Pakistan, in 2015. His research work focuses on phytochemistry and pharmacology, dealing particularly with bio-guided isolation of new compounds for drug discovery. His research interest includes isolation and structure elucidation of bioactive compounds using several innovative techniques followed by in vitro and in vivo biological screening. Dr. Abdur Rauf is the author and coauthor of more than 280 research papers published in peer-reviewed journals (impact factor = 105.94), with 5392 citations as per Google Scholar report. In addition, he has an H-index of 38 and an i10 index of 117 (Google Scholar). He has also written 10 book chapters for international book publishers, and has two national and two international patents for discovery of novel antidiabetic drugs. Furthermore, he is the Associate Editor/Editorial Board Member of *Frontier in Pharmacology, Biocell, Green Processing and Synthesis*, and *Medicinal Chemistry*, among others. He is also a fellow of the American Chemical Society and Bioconlab Oral State University, Russia. He has supervised two PhD and four MPhil students. Dr. Abdur Rauf has won the Young Scientist Award from the Directorate of Science and Technology KP, Pakistan, in the year 2018 and the Research Productivity award for the year 2016–2017 from the Pakistan Science Foundation. He organized (Chief-Organizer) the 1st International Conference on Drug Discovery against Cancer and other Diseases (DDCD) in 2019, and acted as an organizer in the International Pharmacy

Conference and Exhibition on Emerging Trend in Drug Development, Therapeutics, and Commercialization (2019). He has also won eight national and four international projects. Some of his major goals in research include the discovery of eco-friendly and cost-effective protocols for the production of nanoparticles-based new drugs. His research interests also include exploration of the local natural resources (i.e., Pakistani flora) for efficient treatment of different health disorders, which in turn would have a significant impact on the national economy.

About the Editors

Syed Muhammad Mukarram Shah, PhD
Professor in the Department of Pharmacy, University of Swabi, Swabi, Anbar, KP, Pakistan

Syed Muhammad Mukarram Shah, PhD, is a Professor in the Department of Pharmacy, University of Swabi, Swabi, Anbar, KP, Pakistan.

Professor Shah has published more than 40 research papers in various international impact factor journals. His research work is focussed on pharmacological, phytochemicals, and pharmacology, dealing particularly in bio-guided isolation of active compounds for drug discovery. His research interest includes isolation and structure elucidation of bioactive compounds using several innovative techniques followed by in vitro and in vivo biological screening. His research work also explores novel plant-based remedies for the treatment of inflammation, neuroinflammation, Alzheimer's, and depression.

Contents

Contributors

Fazal Akbar
Centre for Biotechnology and Microbiology, University of Swat, Khyber Pakhtunkhwa, Pakistan

Mohammad Ali
Centre for Biotechnology and Microbiology, University of Swat, Khyber Pakhtunkhwa, Pakistan

Zafar Ali
Centre for Biotechnology and Microbiology, University of Swat, Khyber Pakhtunkhwa, Pakistan

Godwin Anywar
Department of Plant Sciences, Microbiology & Biotechnology, College of Natural Sciences, Makerere University, Kampala, Uganda

Maliha Atta
Institute of Chemical Sciences, Gomal University, Dera Ismail Khan, KPK, Pakistan

Yahya S. Al-Awthan
Department of Biology, Faculty of Science, University of Tabuk, Tabuk, Saudi Arabia
Department of Biology, Faculty of Science, Ibb University, Ibb, Yemen

Lubna Azmi
Sitapur Shiksha Sansthan Group of Institutions, Sitapur, Uttar Pradesh, India

Omar Bahattab
Department of Biology, Faculty of Science, University of Tabuk, Tabuk, Saudi Arabia

Nadia Bibi
Department of Microbiology, Shaheed Benazir Bhutto Women University, Peshawar, Khyber Pakhtunkhwa, Pakistan

Mohammed A. Al-Duais
Department of Biochemistry, Faculty of Science, University of Tabuk, Tabuk, Saudi Arabia
Biochemistry Unit, Chemistry Department, Faculty of Science, Ibb University, Ibb, Yemen

Allah Bakhsh Gulshan
Department of Botany, Ghazi University, Dera Ghazi Khan, Pakistan

Zahid Hussain
Centre for Biotechnology and Microbiology, University of Swat, Khyber Pakhtunkhwa, Pakistan

Arshad Iqbal
Centre for Biotechnology and Microbiology, University of Swat, Khyber Pakhtunkhwa, Pakistan

Chenmala Karthika
Department of Pharmaceutics, JSS College of Pharmacy, JSS Academy of Higher Education & Research, Ooty, Tamil Nadu, India

Anees Ahmed Khalil
Department of Chemistry, University of Swabi, Swabi, Anbar, Khyber Pakhtunkhwa, Pakistan

Ammar Ahmad Khan
University Institute of Food Science and Technology, Faculty of Allied Health Sciences, The University of Lahore, Lahore, Pakistan

Muhammad Arslan Khan
Department of Pharmacy, The University of Lahore, Lahore, Pakistan

Muneeb Khan
Riphah College of Rehabilitation and Allied Health Sciences, Riphah International University, Lahore, Pakistan

Aishma Khattak
Department of Bioinformatics, Shaheed Benazir Bhutto Women University Peshawar, Khyber Pakhtunkhwa, Pakistan

Motlalepula Gilbert Matsabisa
Instituto de Quimica de Sao Carlos, Universidade de Sao Paulo, Sao Carlos, Brazil

Abhay Prakash Mishra
Department of Pharmacology, University of Free State, Bloemfontein, South Africa

Neelma Munir
Department of Biotechnology, Lahore College for Women University, Lahore, Pakistan

Saima Naz
Department of Biotechnology, Bacha Khan University Charsadda, Khyber Pakhtunkhwa, Pakistan

Manisha Nigam
Department of Biochemistry, Hemvati Nandan Bahuguna Garhwal University, Srinagar Garhwal, Uttarakhand, India

Uzma Nihar
Department of Zoology, Women University, Mardan, Pakistan

Mayowa Shakirdeen Obidola
Department of Crop Production Technology, Federal College of Forestry, Jos, Nigeria

Ahmed Olatunde
Department of Medical Biochemistry, Abubakar Tafawa Balewa University, Bauchi, Nigeria

Md. Habibur Rahman
Department of Global Medical Science, Yonsei University Wonju College of Medicine, Yonsei University, Wonju, Korea

Abdur Rauf
Department of Chemistry, University of Swabi, Swabi, Anbar, Khyber Pakhtunkhwa, Pakistan

Muhammad Rizwan
Centre for Biotechnology and Microbiology, University of Swat, Khyber Pakhtunkhwa, Pakistan

Muhammad Saleem
Department of Chemistry, Ghazi University, Dera Ghazi Khan, Pakistan

Muddaser Shah
Department of Botany, Abdul Wali Khan University Mardan, Mardan, Pakistan

Nadia Sharif
Department of Zoology, Women University, Mardan, Pakistan

Kishwar Sultana
Centre of Biotechnology and Microbiology, University of Swat, Khyber Pakhtunkhwa, Pakistan

Shrouq H. Sweilam
Department of Pharmacognosy, Faculty of Pharmacy, Prince Sattam Bin Abdulaziz University, Al-Kharj, Saudi Arabia; Department of Pharmacognosy, Faculty of Pharmacy, Egyptian Russian University, Badr City, Cairo, Egypt

Habibu Tijjani
Department of Biochemistry, Bauchi State University, Gadau, Nigeria

Taj Uddin
Centre for Biotechnology and Microbiology, University of Swat, Khyber Pakhtunkhwa, Pakistan

Raquel Peres Morais Urano
Instituto de Quimica de Sao Carlos, Universidade de Sao Paulo, Sao Carlos, Brazil

Archana Yadav
Department of Microbiology, Institute of Biological Sciences and Biotechnology, C. S. J. M. University, Kanpur, Uttar Pradesh, India

Nasib Zaman
Centre for Biotechnology and Microbiology, University of Swat, Khyber Pakhtunkhwa, Pakistan

Abbreviations

AA	antioxidant activity
AAF	allantoic amniotic fluid
AChE	acetylcholinesterase
AMF	*A. marmelos* flower extract
ATP	adenosine triphosphate
AV	avian influenza
BChE	butyrylcholinesterase
C	Clausena
Ca	calcium
EBV	Epstein-Barr virus
EIC	East India Company
EO	essential oil
Fe	iron
HA	hemagglutination
HAS	hydroxyl sanshool
HC	hydrocortisone
HID	hypodermic injections
HPAI	highly pathogenic avian influenza
IFN	interferon
IGA	intragastric administration
IL	interleukin
IMI	intramuscular injections
IPT	intraperitoneal injections
K	potassium
MBC	minimal bactericidal concentration
MDR	multiple drug resistance
Mg	magnesium
MIC	minimal inhibitory concentration
Mn	manganese
MRSA	methicillin-resistant Staphylococcus aureus
Na	sodium

OH•	hydroxyl radicals
ORAC	oxygen radical absorbance capacity
P	phosphorus
RNS	reactive nitrogen species
RSS	reactive sulfur species
STAI	State-Trait Anxiety Inventory
TCM	traditional Chinese medicine
Th1	T helper cell 1
TK lymphoid	thymidine kinase lymphoid
TNF	tumor necrosis factor
URI	upper respiratory infection
UTI	urinary tract infection
VGSCs	voltage-gated sodium channels
Z	*Zanthoxylum*
Zn	zinc

Preface

Medicinal plants are factories that produce valuable compounds. Human beings have used various parts of medicinal plants for curing various diseases for a long time. About 80% of the world population use medicinal plants for basic health needs. The relation among plants, drugs, and humans describe the history of mankind. Medicinal plants produce natural products which have potency to be used in modern medicine for curing various diseases which are not curable. Herbal drugs have received popularity for their traditional uses. Several classes of bioactive compounds that synthesize naturally in plants are known as pharmacological active compounds. The derived bioactive compounds including steroids, tannins, flavonoids, and alkaloids possess several biological applications. Various techniques are reported for the isolation and extraction of bioactive compounds and essential oils from medicinal plants.

This book aims to document treatments with herbal phytochemicals for various diseases. This book titled *Phytochemical and Pharmacological Investigation of the Family* Rutaceae focuses on: Part I: Introduction and usage of Family Rutaceae; Part II: Phytochemical and pharmacological studies of Family Rutaceae; Part III: Nutritional clinical and toxicological profile of Family Rutaceae.

The Rutaceae family comprises numerous ornamental as well as economically significant fruit-bearing trees. Family members of Rutaceae have separate oil glands that bear aromatic leaves and their flowers are mostly perfect but can be unisexual occasionally. This family comprises citrus spp. (lime, lemon, mandarin, sweet orange, and grapefruits), Aegle spp., Atalantia spp., Clausena spp., Lunasia, spp., etc. All the members of the family Rutaceae are known for their phytochemical composition and associated medicinal properties.

The information documented in the 11 chapters in this volume are important for new drug discovery and, as a result, for the pharmaceutical industries. These book chapters can also direct researchers to isolate/ extract pure compounds from the Family Rutaceae for new and novel drug discovery.

—Abdur Rauf, PhD
Editor

CHAPTER 1

The Family Rutaceae: An Introduction

ANEES AHMED KHALIL[1], MUNEEB KHAN[2], ABDUR RAUF[3], YAHYA S. AL-AWTHAN[4,5], and OMAR BAHATTAB[4]

[1]*University Institute of Diet and Nutritional Sciences, Faculty of Allied Health Sciences, The University of Lahore, Lahore, Pakistan*

[2]*Riphah College of Rehabilitation and Allied Health Sciences, Riphah International University, Lahore, Pakistan*

[3]*Department of Chemistry, University of Swabi, Swabi, Anbar, Khyber Pakhtunkhwa (KP), Pakistan*

[4]*Department of Biology, Faculty of Science, University of Tabuk, Tabuk, Saudi Arabia*

[5]*Department of Biology, Faculty of Science, Ibb University, Ibb, Yemen*

ABSTRACT

Family Rutaceae is a family mainly comprising of flowering plants and is placed under the order Sapindales. It has been reported to encompass several species (2070) and genera (150–170). All the plants in family Rutaceae are shrubs/trees and their distribution is found across the globe specifically in tropical and warm temperate regions. Rutaceae family contains numerous ornamental as well as economically significant fruit bearing trees. Family members of Rutaceae have distinct oil glands bearing aromatic leaves and mostly their flowers are perfect but they can be unisexual occasionally.

Phytochemical and Pharmacological Investigation of the Family Rutaceae.
Abdur Rauf, Hafiz Ansar Rasul Suleria, & Syed Muhammad Mukarram Shah (Eds.)

This family contains citrus spp. (lime, lemon, mandarin, sweet orange, and grapefruits), Aegle spp., Atalantia spp., Clausena spp., Lunasia, spp., etc. All the members of family Rutaceae are known for their phytochemical composition and associated medicinal properties. This chapter provides an overview regarding some main genus and species present in family Rutaceae.

1.1 INTRODUCTION

Rutaceae is considered to be a large family containing various shrubs, trees, and herbs of nearly 1500–2040 species in 150–170 genera.[12] This family is widely distributed around the globe specifically in different regions of South Africa, Australia, and Asia. Species present in family Rutaceae are known to produce essential oils, linonoids, coumarins, flavonoids, and several alkaloids such as acridone and carbazole. Botanically, Rutaceae are mainly characterized by three features, that is, leaves having clear dots with laminating oil cells; pure white having visibly oily stigma; and fruits are berry shaped and fleshy.[3] Flowers of Rutaceae are classified as perfect (unisexual); sepals are in connation or distinct and the number can vary from 2 to 5; petals are 2–5 in number and there length ranges from few millimeters to centimeters; stamens vary from 2 to 60 in number; and filaments are distinctive while they can be sometimes connate.[3]

Plants in Rutaceae have been known for their industrial beneficial role such as oil extracted from peels of sweet orange (*Citrus sinesis*) which are used in perfumery industry as a natural flavoring agent.[1] Around the globe, the most known commodity owing to its juice and Vitamin C content are citrus fruits. Members of Rutaceae family encompass abundant source of Vitamin C and various alkaloids.[25] *Citrus bergamia* (Bergamot orange) peel oil is also used in perfumery and medicine industry. Similarly, oil mechanically extracted from peels of lemon and Citrus aurantium is considered carminative and a natural flavoring agent. Family Rutaceae are well renowned among scientific community owing to their health-promoting benefits, that is, species of genus *Pilocarpus*, *Agathosma*, *Ruta*, and *Zanthoxylum* consists of pilocarpine (imidazole alkaloid) that has been used for the treatment of glaucoma. Different parts (oil, leaves, seeds, bark, etc.) of various species present in genus *Ruta*, *Agathosma*, *Murraya*, and *Zanthoxylum* are used for promoting menses and urination, and to stop spasms and flatulence.[34] Interestingly, plants classified in the Rutaceae family comprise unique alkaloids like acridone, quinazoline, and quinolone that possess antibacterial, cytotoxic, and neuroprotective properties. Evidently, coumarins present in the Rutaceae family are also known for their antitumor properties. As mentioned earlier, family Rutaceae comprises up to 2040 species so it is not possible to discuss all in detail. However, herein this chapter we shall discuss nearly ten genus of Rutaceae family.[20]

1.2 GENUS AEGLE

Genus *Aegle* belongs to family Rutaceae and comprise three accepted species namely *Aegle decandra* Fern.-Vill., *Aegle glutinosa* (Blanco) Merr., and *Aegle marmelos* (L.) Corrêa. Out of these three, most studied specie is *Aegle marmelos* (L.) Corrêa. as shown in Figure 1.1. It is commonly known as Bel/Bael (Hindi), Shivadruma (Sanskrit), and wood apple/golden apple (English) and is predominantly known for its medicinal properties in Asia.[14]

FIGURE 1.1 *Aegle marmelos (L.) Corrêa.*

Source: Image by Vijayanrajapuram. https://creativecommons.org/licenses/by-sa/4.0/

Indigenously, bael trees are predominantly cultivated in Himalayan regions, Pakistan, India, Cambodia, Bangladesh, Sri Lanka, Nepal, and Vietnam. In India, leaves of bael tress are quite often offered to various Indian gods such as Lord Shiva; hence, this plant is abundantly available in premises of different temples.[6,19] *Aegle marmelos* (L.) is known as a subtropical tree that belongs to genus Aegle having slow growth rate with a medium size height. Annually, during summer season (March–June), this tree bears fruit, which in shape varies from round to oblong and possesses health-promoting benefits. Ripe fruits of Bael are used for the preparation of "sherbet" and are known for its cooling effects. But pulp of Bael is processed to form syrups, murabba, and marmalade and consumed with indigenous breads.[19]

Depending upon organoleptic properties and fruit size, Bael is categorized into different varieties, that is, *gonda* (1, 2, and 3), *Ojha, Mitzapuri, Kaghzi gonad, Baghael,* and *Azamati*. Fruit pulp of Bael has a cache of vitamins (vitamin A, niacin, Vitamin C, riboflavin, and thiamine) and minerals (phosphorous and calcium).[21,26] Bael tree has reported to grow significantly without any retardation in growth at an elevated altitude (1200 m) and in a temperature range of 50 to −7°C. Bael tree is a deciduous plant that grows up to 12 m in height and 1.20 m in diameter. Trunk of Bael tree is thick, and short having narrow oval-shaped ends. Bael tree branches are small and nonspiny while bark is soft and bluish-gray in color. The leaves of Bael are alternate, deciduous, aromatic, and trifoliolate; however, the leaf petiole are 4.2–6.7 cm in length. Flower comprises 4–5 petals along with fifty or more stamens that are green-yellowish in color. Bael fruit is 5 to 20 cm in diameter having hard wooden pericarp, while the seeds are 1 cm long in oblong shape.[22]

1.3 GENUS ATALANTIA

Another genus that belongs to family Rutaceae is known as genus *Atalantia,* which comprises nearly 71 species. Out of these, 18 species are classified as accepted, that is, *Atalantia armata (Thwaites)* Guillaumin, *Atalantia buxifolia (Poir.)* Oliv., *Atalantia ceylanica (Arn.)* Oliv., *Atalantia citroides Pierre ex* Guillaumin, *Atalantia dasycarpa* C.C. Huang, *Atalantia henryi (Swingle)* C.C. Huang, *Atalantia macrophylla (*Oliv.*)* Kurz, *Atalantia guillauminii Swingle, Atalantia monophylla* DC., *Atalantia hainanensis* Merr. & Chun ex Swingle, *Atalantia racemosa* Wight ex Hook., *Atalantia wightii* Yu. Tanaka, *Atalantia roxburghiana var. kwangtungensis (Merr.)* Swingle, *Atalantia stenocarpa* Drake, *Atalantia roxburghiana* Hook.*f.*, *Atalantia simplicifolia (Roxb.)* Engl., and *Atalantia sessiliflora* Guillaumin. Among these, most studied and investigated species are *Atalantia monophylla* DC, and *Atalantia roxburghiana* Hook.f. Figure 1.2 shows the images of some of these species.

FIGURE 1.2 Images of species in genus *Atalantia.*

Genus *Atalantia* Correa is known to possess eleven resembling species that are naturally grown in various countries like India, Pakistan, Southern China, Srilanka, Vietnam (North and South), Thailand, Cambodia, Myanmar, Peninsular Malaya, Sumatra, and Java. Generally, the plants in this genus resemble with *citrus* fruit tree and also bear small round orange/lime like fruits that are green-yellowish in color. These plants are small medium ranged and are spinous shrubs. Species present in this genus are abundant in alkaloids, essential oils, and flavonoids, which possess bioactive properties. These plants are helpful in treating paralysis, rheumatoid arthritis, bacterial, and fungal infections.[13] Plants in *Atalantia* are classified as closely wild relative of citrus and therefore these species are used as a rootstock for commercial grafting of citrus fruit trees. In India, four species of genus *Atalantia* are well known, that is, *A. monophylla* (L.) DC., *A. racemosa* Wight var. *racemose, A. wightii* Tanaka, and *A. simplicifolia* (Roxb.) Engl. Among these, the most rare species is *A. simplicifolia* as its occurrence is found in very few areas like Nagaland, Assam, Mizoram, and Meghalaya.[29]

A. monophylla (L), a well-renowned species of genus *Atalantia,* is locally known as Kattuelumichai, Nimbu Bannimbu, Kattunarakam, and Jungli Nimbu Banjamir.[10] Different parts of *A. monophylla* possess medicinal properties like oil extracted from its seeds which are helpful in treating arthritis.[13] *Atalantia monophylla* DC. is a shrub having a height of nearly 6 m and is abundantly cultivated in dry, sandy, and rocky areas of Southeastern Asia. The pedicel of fruit is one and a half centimeter in length with the fruit being 2 cm in diameter. The dimensions of its leaf blade and petiole are 9.4×6 cm and 1.4 cm, respectively. The flower of this plant is 1.5 cm in length. On the other hand, *A. simplicifolia* and *A. roxburghiana* Hook.f. are small shrubs/treelets having height of up to 5–6 m. Stems of *A. roxburghiana* Hook.f. are spiny while the flower is white in color having four lobed calyx with free stamens. Fruits are pale yellow in color and leaves are spiral. Flowers of *A. simplicifolia* are small having 4 petals that are obovate-oblong and 8 free stamens.[34]

1.4 GENUS CITRUS

Genus *Citrus* comprises flowering shrubs/trees that belong to family Rutaceae. According to The Plant list, nearly 392 species are present in this genus but only 33 of these are known as accepted species. Some of the world most renowned fruits that belong to genus citrus are Pumelos (*C. maxima*), Grapefruit (*C. paradisi*), Lemons (*C. limon*), Limes (*C. aurantiifolia and C. larifolia*), Oranges (*C. sinensis*), and Mandarins (*C. reticulata*). Genus *Citrus* is known to be native to Asia; however, it is extensively grown in Australia and Melanesia. Since ancient times, numerous citrus species are cultivated indigenously in these areas. Later, cultivation of citrus spread to Mediterranean and Middle East due to trade route and further to Central Europe. Similarly, Citrus is being cultivated in Polynesia and Micronesia owing to Austronesian expansion.[15,37] Majority of Citrus species cultivated around the globe are either natural or artificial hybrids of certain ancient species, including papeda, mandarin, and pomelo.[11] Some of the plants in Genus *Citrus* are shown in Figure 1.3.

FIGURE 1.3 Images of species in genus *Citrus*.

C. hystrix DC. is a shrub having a height up to 2 m. The bark of this plant is green-yellowish in color with a smooth texture while stem and leaves are light to dark green in color. The blade of leaves contains spotted oil cells and fruit is green in color with a strong aroma. Another specie belonging to genus *Citrus* is *C. mitis* Blco. (*Citrus microcarpa* Bunge) which is commonly an ornamental treelet in China and its availability is known in Philippines and Indonesia since ancient times. The height of this plant is up to 3 m and fruit is yellowish orange in color having dimensions of 2 × 3.5 cm. Likewise, *C. reticulate* (mandarin) is a type of citrus fruit tree. Mandarins are small, oblate pebbly-skinned, and tender. These are grown in subtropical and tropical areas.[2] Height of mandarin tree is up to 7.6 m and this tree has thorny branches and trunk. Mature tree yields up to 79 kg of fruit.[2]

1.5 GENUS CLAUSENA

Genus *Clausena* is a flowering tree that belongs to family Rutaceae. In 1768, a Dutch botanist Nicolaas Laurens Burman was the first person to define this plant. It is predominantly cultivated in Asia, Pacific Islands, Australia, and Africa. This genus comprises 21 accepted species namely *Clausena abyssinica* Engl., *Clausena dentata (Willd.)* Roem., *Clausena anisata* (Willd.) Hook.f. ex Benth., *Clausena engleri* Yu. Tanaka, *Clausena brevistyla* Oliv., *Clausena excavata* Burm.f., *Clausena austroindica* B.C. Stone and K.K.N. Nair, *Clausena hainanensis* C.C. Huang and W.F. Xing, *Clausena indica* (Dalzell) Oliv., *Clausena harmandiana* (Pierre) Guillaumin, *Clausena inolida* Z.J. Yu and C.Y. Wong, *Clausena henryi* (Swingle) C.C. Huang, *Clausena kanpurensis* Molino, *Clausena heptaphylla* (Roxb.) Wight and Arn., *Clausena lansium* (Lour.) Skeels, *Clausena poilanei* Molino, *Clausena lenis* Drake, *Clausena wallichii* Oliv., *Clausena luxurians* (Kurz) Swingle, *Clausena smyrelliana* P.I.Forst., and *Clausena sanki* (Perr.) Molino. Figure 1.4 shows some of species present in this genus.

FIGURE 1.4 Images of species in genus *Clausena*.

C. excavata Burm. f. is a shrub grown in different regions of Philippines and India. This plant is wild and cultivated in villages, forests edges, and lowland. Leaves of this plant are spiral in shape along with asymmetrical folioles (5–31 pairs). The inernodes and rachis are 1.5 cm and 29 cm in length. Flowers of this plant are small and covered in panicles that are 15 cm long. *C. excavata* fruit is 5 mm in diameter on pedicels that are 7 mm in length. Another plant in this genus is *C. lansium* (Lour.) Skeells, which is a shrub having a height of nearly 9 m. Leaves of this plant are spiral with three to four asymmetric folioles and rachis are 15 cm in length. Ripe fruits are sour in taste with golden-yellow colored skin and white pulp that surrounds numerous seeds (green-blackish). Likewise, *C. anisata* (Willd.) Hook.f. ex Benth. is abundantly grown in sub-Saharan Africa and is a deciduous shrub that belongs to family Rutaceae. Further, it is cultivated in India, Pakistan, South-east Asia, Queensland, Indonesia, and Malaysia. Another shrub, that is, *C. smyrelliana* having a height of 4 m belongs to family Rutaceae.

1.6 GENUS TETRADIUM

Genus *Tetradium* L. belongs to family Rutaceae and in old books, certain species now included in *Tetradium* were classified under genus *Euodia.* Species in this genus are available in temperate to tropical Eastern Asia. According to The Plant List eight species in this genus are accepted while seven of these are commonly cultivated in Northern China.[7] The names of these species are *Tetradium austrosinense* (Hand.-Mazz.) T.G. Hartley, *Tetradium trichotomum* Lour., *Tetradium calcicolum* (Chun ex C.C. Huang) T.G. Hartley, *Tetradium sambucinum* (Blume) T.G. Hartley, *Tetradium daniellii* (Benn.) T.G. Hartley, *Tetradium ruticarpum* (A.Juss.) T.G. Hartley, *Tetradium glabrifolium* (Champ. ex Benth.) T.G. Hartley, and *Tetradium fraxinifolium* (Hook. f.) T.G. Hartley. Since ancient times, *Tetradium ruticarpum* has been used as a folk medicine in Chinese culture and hence has been studied extensively. Unripe *T. ruticarpum* fruit is present in various versions of Chinese Pharmacopoeia; similarly, in Korea and Japan this herb has been found to prescribe as traditional medicine. Traditionally in China, *T. ruticarpum* was used for curing emesis, epigastric pain, headache, and aphtha. Figure 1.5 shows some species present in this genus.[7]

FIGURE 1.5 Images of species in genus *Tetradium.*

Plants in genus Tetradium attain a height of 3–5 m and are deciduous/evergreen shrubs. Their leaves are odd-pinnate and flowers are dioecious having 4–5 petals, stamens, sepals, and carpels. Fruits of these plants have a connate base and comprise one to five follicles. They have cartilaginous endocarp and dry exocarp. Their flowering season is from April to September while their fruiting stage is in August to November.[36]

1.7 GENUS LUNASIA

Species of Genus *Lunasia* are abundantly present in Northern Queensland, Philippines, East Java, and New Guinea. The plants in this genus are known as small shrubs that are evergreen and deciduous. Two reported varieties of *L. amara* Blanco are babuyanica (Merr.) Hartley and amara. Figure 1.6 shows the image of *Lunasia amara.*

FIGURE 1.6 Image of *Lunasia amara.*

Image credit: Reprinted from https://toptropicals.com/catalog/uid/lunasia_amara.htm

The origin of *L. Blanco* is Malaysian but also it is grown in Philippines, Java, South Borneo, Guinea, and Australia (Cape York Peninsula). This genus is quite different from some other genera present in family Rutaceae due to its small and head-like clustered trimerous flowers. Certain species such as *L. obtusifolia, L. quercifolia, L. parvifolia,* and *L. mollis* differ from *L. amara* due to their vegetative characteristics. These species usually thrive well in lowlands, well-drained rainy forests, dry thickets, and regrown gardens. Same ecological characteristics are observed in various Malaysian plant species.[8] *L. amara* Blanco is a small tree or shrub having height of nearly 12 m that is sparsely branched. At brenchlet end, the leaves of these plants are crowded on 1.5–15 cm long petioles and blades are 5.5–60 cm in length. Species of this genus have smooth twigs and flowers are clustered having 3–6 mm in diameter. They possess green-yellowish petals and 0.5-mm-long sepals. Seeds are dark to red-brownish in color. This plant flowers and bears fruits all around the year. They are mostly cultivated via seed propagation.[18]

1.8 GENUS MICROMELUM

Genus *Micromelum* belongs to family Rutaceae and comprises nearly nine different species. These species are widespread in Pakistan, India, Australia, South China, Fiji, and Samoa.[28,33] Different species of genus *Micromelum* are a rich source of coumarins specifically 6-coumarin and 8-prenylated coumarin. Additionally, they possess derivatives of dihydrocinnamic acid, and carbazole alkaloids.[23,31] Various parts of species present in genus *Micromelum* are an abundant source of alkaloids, polyoxygenated flavonoids, and coumarins.[27,30] Figure 1.7 shows the images of some of the most investigated species of genus *Micromelum.*

FIGURE 1.7 Images of species in genus *Micromelum.*

M. integerrimum is a small tree that attains a height of up to 8 m and is usually grown in sandy soil and moist mountainous forest ranges. It is most widely available from sea level to as high as 2000 m in Vietnam, Nepal, Bhutan, China, India, and Cambodia.[35] This specie is known for abundance of secondary metabolites such as coumarin derivatives. These coumarin derivatives isolated from this specie have shown significant bioactivities like antiplatelet, cytotoxic, and antimutagenic properties.[23] Height of *M. minutum*, a small spineless tree, is reported to be as high as 3 m. It is extensively cultivated in Pacific islands and Southeastern Asia. Various parts of this plant like roots, flower, leaves, and fruit have reported pharmacological properties.[16,30] Several studies have reported presence of alkaloids, terpenes, coumarins, and phenylpropanoids in different parts of this plant.[9] Another specie present in this genus is *M. falcatum*. This small tree attains a height of 1–3 m. Flowering season of this plant is from January to April and is found in mountainous areas in Vietnam, China, Myanmar, Cambodia, and Thialand.[24]

1.9 GENUS PARAMIGNYA

Another genus in family Rutaceae is *Paramignya*, which is abundantly grown in Australia, Thailand, South Vietnam, Sri Lanka, and Philippines. It generally comprises twenty species while among them just two are accepted according to The Plant List.[4,5] Species in this genus are well known owing to their traditional use as medicinal agents. In addition, morphologically this genus resembles genus Luvunga as it contains six to ten stamens, three to five petals, and two to four locules in ovary. Leaves of this genus are smaller as compared to common trifoliolate leaves.[17]

KEYWORDS

- **Rutaceae**
- **shrubs/trees**
- **fruit-bearing**
- **unisexual**
- **aromatic leaves**
- **phytochemical composition**

REFERENCES

1. Ahmed, Z. U.; M. A. Hassan; Z. N. T. Begum; Khondker, M.; Kabir, S. M. H.; Ahmad, M.; Ahmed, A. T. A. Eds.; In *Encyclopedia of Flora and Fauna of Bangladesh. Angiosperms: Dicotyledons*; Asiatic Society of Bangladesh: Dhaka, 2009; vol 10, pp 159–187.
2. Akter, A.; Nadarajan, J. In *Optimization of Cryopreservation Protocols for Zygotic Embryos of Citrus Reticulata*, III International Symposium on Plant Cryopreservation, 2018; vol 1234, pp 137–144.
3. Chase, M. W.; Morton, C. M.; Kallunki, J. A. Phylogenetic Relationships of Rutaceae: A Cladistic Analysis of the Subfamilies using Evidence from RBC and ATP Sequence Variation. *Am. J. Bot.* **1999,** *86* (8), 1191–1199.
4. Cuong, N. M.; Huong, T. T.; Khanh, P. N.; Van Tai, N.; Ha, V. T.; Tai, B. H.; Kim, Y. H. Paratrimerins A and B, Two New Dimeric Monoterpene-Linked Coumarin Glycosides from the Roots and Stems of Paramignya trimera. *Chem. Pharm. Bull.* **2015,** *63* (11), 945–949.
5. Dang, P. H.; Le, T. H.; Phan, P. K.; Le, P. T.; Nguyen, M. T.; Nguyen, N. T. Two Acridones and Two Coumarins from the Roots of Paramignya trimera. *Tetrahedron Lett.* **2017,** *58* (16), 1553–1557.
6. Das, B.; Das, R. Medicinal Properties and Chemical Constituents of Aegle marmelos Correa. *Indian Drugs* **1995,** *32*, 93–99.
7. Flora of China Editorial Committee. In *Flora of China*; Science Press: Beijing, China, 2006.
8. Hartley, T. G. A Revision of the genus Lunasia (Rutaceae). *J. Arnold Arbor.* **1967,** *48*, 400–475.
9. Hasan, C. M.; Ahsan, M. Chemical Constituents. Cytotoxic Activities and Traditional uses of Micromelum minutum (Rutaceae): A Review. *Pharm. Pharmacol. Int. J.* **2019,** *7* (5), 229–236.
10. Kandappa, H. R.. In Vitro Antifungal, Antioxidant and Cytotoxic Activities of a Partially Purified Protein Fraction from Atlantia monophylla Linn (Rutaceae), Leaf Trop. *J. Pharm. Res.* **2015,** *14 (3)*, 487–493.
11. Klein, J. D. Citron Cultivation, Production and uses in the Mediterranean Region. In *Medicinal and Aromatic Plants of the Middle-East*; Springer: Dordrecht, 2014; pp 199–214.
12. Kubitzki, K.; Kallunki, J. A.; Duretto, M.; Wilson, P. G. Rutaceae. In *The Families and Genera of Vascular Plants*; Kubitzki, K., Ed.; Springer: Berlin, 2011; vol 10, pp 276–356.
13. Kumar, R. B.; Narayana, B. S. Tribal Medicinal Studies on Sriharikota is Land, Andhra Pradesh. *Ethnobot. Leafl.* **2010,** *14*, 95–107.
14. Ladaniya, M. S. Commercial Fresh Citrus Cultivars and Producing Countries. In *Citrus Fruit: Biology, Technology and Evaluation*; Academic Press: San Diego, 2008; pp 13–65.
15. Langgut, D. The Citrus Route Revealed: From Southeast Asia into the Mediterranean. *HortScience* **2017,** *52* (6), 814–822.

16. Lekphrom, R.; Kanokmedhakul, K.; Sangsopha, W.; Kanokmedhakul, S. A New Coumarin from the Roots of *Micromelum minutum*. *Nat. Prod. Res.* **2016,** *30* (21), 2383–2388.
17. Mabberley, D. J. Australian Citreae with Notes on Other Aurantioideae (Rutaceae). *Telopea* **1998,** *7* (4), 333–344.
18. Macabeo, A. P.; Aguinaldo, A. Chemical and Phytomedicinal Investigations in Lunasia amara. *Phcog. Rev.* **2008,** *2* (4), 317.
19. Maity, P.; Hansda, D.; Bandyopadhyay, U.; Mishra, D. K. Biological Activities of Crude Extracts and Chemical Constituents of Bael Aegle marmelos (L.) Corr. *Indian J. Exp. Biol.* **2009,** *47*, 849–861.
20. Michael, J. P. Quinoline, Quinazoline and Acridone Alkaloids. *Nat. Prod. Res.* **2004,** *21*, 650.
21. Parmar, C.; Kaushal, M. K. In *Wild Fruits of the Sub-Himalayan Region*; Kalyani Publishers: New Delhi, 1982.
22. Pathirana, C. K.; Madhujith, T.; Eeswara, J. Bael (Aegle marmelos L. Corrêa), a Medicinal Tree with Immense Economic Potentials. *Adv. Agric.* **2020,** *2020*.
23. Phakhodee, W.; Pattarawarapan, M.; Pongparn, P.; Laphookhieo, S. Naturally Occurring Prenylated Coumarins from *Micromelum integerrimum* Twigs. *Phytochem. Lett.* **2014,** *7*, 165–168.
24. Ramesh, B.; Pugalendi, K. V. Antioxidant Role of Umbelliferone in STZ-Diabetic Rats. *Life Sci.* **2006,** *79* (3), 306–310.
25. Roy, D.; Rahman, A. H. M. M. Systematic Study and Medicinal uses of Rutaceae Family of Rajshahi District, Bangladesh. *Plant Environ. Dev.* **2016,** *5* (1), 26–32.
26. Roy, S. K.; Khurdiya, D. S. Other Subtropical Fruit. In In *Handbook of Fruit Science and Technology: Production Composition Storage and Processing*; Salunkhe, D. K., Kadam, S. S., Eds.; Marcel Dekker: New York, 1995.
27. Sahakitpichan, P.; Tanpatanan, W.; Chimnoi, N.; Ruchirawat, S.; Kanchanapoom, T. Glucosides of Phenylpropanoic Acid Derivatives and Coumarins from *Micromelum minutum*. *Phytochem. Lett.* **2015,** *14*, 12–16.
28. Singh, R. K. Lectotypification of the Name *Micromelum integerrimum* (Rutaceae). *Phytotaxa* **2016,** *255* (2), 172–174.
29. Stone, B. C. Rutaceae. In *A Revised Handbook to the Flora of Ceylon*; Dassanayake, M. D., Fosberg, F. R., Eds.; Amerind: New Delhi, 1985; vol 5, pp 176–406.
30. Susidarti, R. A. Chemical Constituents and Biological Activities of *Micromelum Minutum* (Rutaceae) and Two Eugenia Species (Myrtaceae). Ph.D. Thesis, Universiti Putra Malaysia, 2003.
31. Suthiwong, J.; Sriphana, U.; Thongsri, Y.; Promsuwan, P.; Prariyachatigul, C.; Yenjai, C. Coumarinoids from the Fruits of *Micromelum falcatum*. *Fitoterapia* **2014,** *94*, 134–141.
32. Tantishaiyakul, V.; Pummangura, S.; Chaichantipyuth, C.; Ma, W. W.; McLaughlin, J. L. Phebalosin from the Bark of Micromelum minutum. *J. Nat. Prod.* **1986,** *49* (1), 180–181.
33. Vidar, W. S.; Macabeo, A. P. G.; Knorn, M.; Kohls, P.; Aguinaldo, A. M. Polymethoxylated Flavones from *Micromelum compressum*. *Biochem. Syst. Ecol.* **2013,** *50*, 48–50.

34. Wiart, C. Medicinal Plants Classified in the Family Rutaceae. In *Medicinal Plants of Asia and the Pacific*; CRC Press, 2006; pp 211–231.
35. Zhang, D.; Hartley, T. G.; Mabberley, D. J. In *Flora of China*; FOC, 1998; vol 11, p 51.
36. Zhao, Z.; He, X.; Han, W.; Chen, X.; Liu, P.; Zhao, X.; Wang, X.; Zhang, L.; Wu, S.; Zheng, X. Genus Tetradium L.: A Comprehensive Review on Traditional uses, Phytochemistry, and Pharmacological Activities. *J. Ethnopharmacol.* **2019,** *1*, 337–354.
37. Wu, G. A.; Terol, J.; Ibanez, V.; López-García, A.; Pérez-Román, E.; Borredá, C.; ... & Talon, M. Genomics of the Origin and Evolution of Citrus. *Nature* **2018,** *554* (7692), 311–316.

CHAPTER 2

The Family Rutaceae: An Overview of Its Traditional Uses

ANEES AHMED KHALIL[1], MUNEEB KHAN[2], ABDUR RAUF[3], SAIMA NAZ[4], YAHYA S. AL-AWTHAN[4,5], and OMAR BAHATTAB[4]

[1]*University Institute of Diet and Nutritional Sciences, Faculty of Allied Health Sciences, The University of Lahore, Lahore, Pakistan*

[2]*Riphah College of Rehabilitation and Allied Health Sciences, Riphah International University, Lahore, Pakistan*

[3]*Department of Chemistry, University of Swabi, Swabi, Anbar, Khyber Pakhtunkhwa (KP), Pakistan*

[4]*Department of Biology, Faculty of Science, University of Tabuk, Tabuk, Saudi Arabia*

[5]*Department of Biology, Faculty of Science, Ibb University, Ibb, Yemen*

ABSTRACT

Rutaceae family comprises various genera and species that are all mainly flowering plants. All the plants present in the Rutaceae family are well distributed around the world and have been used by diversified population owing to their medicinal properties. Predominant genera present in the family Rutaceae such as *Citrus, Aegle, Zanthoxylum, Micromelum, Atalantia, Lunasia, Clausena, Limonia, Murraya,* and *Paramignya* are being utilized traditionally by people living in different regions of the

Phytochemical and Pharmacological Investigation of the Family Rutaceae.
Abdur Rauf, Hafiz Ansar Rasul Suleria, & Syed Muhammad Mukarram Shah (Eds.)

world due to their medicinal properties. Ayurvedic, Unani, and traditional Chinese medication systems have revealed the usage of various plants of Rutaceae family for the treatment of headaches, stomachaches, vomits, infertility, malaria, skin ailments, paralysis, inflammation, snakebites, hemiplegia, nausea, bloating, and itches. This chapter basically highlights the traditional uses of some important genus of Rutaceae family. This information regarding traditional medicinal properties of different plants of Rutaceae may be helpful for pharmaceutical sector.

2.1 INTRODUCTION

Since ancient times various plants were used for the treatment of different diseases. Several medicinal plants being prescribed by traditional practitioners aided community in curtailing the burden of different ailments. Traditionally, owing to the lack of modern facilities, practitioners at that time lacked the knowledge regarding probable mechanism of action of these medicinal plants. Contemporary sciences have highlighted and validated the mode of action of metabolites present in medicinal plants that have been known since ancient times. Worldwide, plants are known to be a significant source of medicine and therefore play a vital role in human wellbeing. Various countries of the world nearly two-thirds of the world's habitat are dependent on medicinal plants for their basic health care since ancient times. In Asia, numerous medicinal and herbal plants are used for curing different metabolic syndromes.[65] Plants provide ample amount of nutrients, and phyto-nutrients for the effective management of various health conditions. Majority of world's population are of the view that plant and their derived metabolites are safe in comparison to synthetically developed drugs. Additionally, different types of resistant infections have also been less resistant owing to the use of medicinal plants.[46] Keeping in view the importance of medicinal plants since ancient times, this chapter focuses on traditional uses of some plants of Rutaceae family.

2.2 TRADITIONAL USES OF AEGLE

Aegle marmelos (L.) have been used since ancient times to treat constipation as this plant has found to tone up and clean intestines. For optimum results, ripened fruit pulp has been used for the preparation of a traditional syrup locally known as "Sherbet." To treat dysentery and chronic diarrhea, unripe or partially ripened *Aegle marmelos* (L.) fruit is considered the most efficient remedy. Likewise, fusion of *Aegle marmelos* (L.) leaves was considered a significant candidate for the treatment of peptic ulcers. In home remedies, roots of Bael tree are used for the treatment of ear problems. Bael tree leaves are used for the preparation of a medicated oil that helps in providing relief from recurring colds.[53] Burmese, the people living in Myanmar, uses the juice extracted from young leaves of bael tree for the treatment of ophthalmia. Similarly, the people of Indonesia have been consuming ripe bael fruit for the promotion of digestion and soothing of rectum inflammation; mature leaves were applied externally for deflating swellings, sores, and inflammation. Drink prepared from the infusion of bael roots has been ingested to ease palpitation of heart. Malaysian people utilize decocted drink of Bael for stopping the vomit; however, Cambodians used bael fruits for treating liver-related diseases and tuberculosis. Further, Indians recognize this plant as sacred and use its leaves for worshiping their Gods like Shiva. While in Sri Lanka, this plant is well known for treating diabetes since ancient times.[69]

Pulp of Bael fruit is satisfyingly sweet in taste and is well flavored. Locally, pulp of this fruit is employed as a well-renowned summer drink in India and Pakistan. Purposely, the soft textured pulp of this fruit is scooped and seeds are removed and later blended in milk with the addition of cardamom and sugar: this prepared drink has been used in summer for its cooling effects. Semi-mature fruit of Bael is processed for the preparation of jam through the addition of preservatives, citric acid, and sugar. In some parts of Asia the dry pulp of Bael is also consumed locally. Other than its dietary consumption, in Ayurvedic medications, traditionally the Bael is used for treating diarrhea. Unripen Bael fruit is used to prevent stomach irritation and helps in aiding digestion system. Semi-ripen bael fruit has been used as a promising antidiarrheal agent and improves digestive system's functionality. Fully matured Bael fruit is considered to be more helpful as compared to immature raw fruit and has been used for preventing chronic and subacute dysentery. Leaves of Bael tree are used traditionally for treating stomach pain, vomiting, abdominal pains, fever, heart palpitation, dyspepsia, and seminal weakness.[6,18,52]

In Unani, Ayurvedic, and Siddha ancient medications systems being practiced in India, *A. marmelos* has been prescribed as a local medicinal agent for the treatment of diabetes. Furthermore, traditionally different parts of Bael like fruit, roots, seeds, pulp, and stems have been used for the treatment of epilepsy, inflammation, fever, constipation, acidity, jaundice, eye disorders, tumors, dysentery, myalgia, bronchitis, diarrhea, ulcers, stomachaches, nausea, mental illness, snakebites, spermatorrhea, febrile delirium, small pox, burning sensations, and upper respiratory tract infections. Likewise, it is used in treating swollen joints, diarrhea, wounds, fractures, hypertension, and anemia. Sweetened drink formulated from Bael fruit pulp has an easing effect on patients who have recently recovered from bacillary dysentery. Bael oil was used for minimizing the peculiar burned sensations occurring in the soles. Decocted bark and roots of Bael tree help in reducing fever, while the poultice of leaves has been used for treating opthalmia.[31,32,41,48]

In the Ayurvedic medicinal system of India, *A. marmelos* have attained a predominant position among other medicinal plants. It is being described as symbol healing tree and fertility, which strengthens the human body. This recognition is mainly due to the medicinal perspectives of different parts of Bael tree, that is, seeds, fruit, leaf, root, and bark that have been used for curing different syndromes. As a matter of fact, according to Charaka (1500 BC), a well-ranked Ayurvedic priest states that no drug is better known to people living in India than *Aegle marmelos*. It has been used alone and/or in combination with other medicinal formulations for the treatment of a wide range of diseases. In the Indian subcontinent, various parts of *Aegle marmelos* such as whole fruit, seeds, leaves, bark, and roots are valued as a potential herbal plant for the treatment of diseases. It is known that leaves of Bael possess expectorant potential and therefore are helpful in managing piles, bowel complaints, diarrhea, and dropsy more effectively. However, its roots are proposed to be effectual in treating abdominal pains, heart palpitation, fevers, flatulence, dysentery, colitis, and urinary problems. Ripen bael fruit significantly acts as a laxative.[17,43]

2.3 TRADITIONAL USES OF GENUS ZANTHOXYLUM

Various species of genus Zanthoxylum have been traditionally used for the treatment of different diseases. Among these, bark of *Zanthoxylum Avicennae* (Lamk.) DC has been used by people living in Vietnam, Laos, and Cambodia as a bitter tonic. But, Philippines use a drink formulated by boiling stems of *Z. Avicennae* that helps in promoting digestion and mitigates the snakebite poisoning. Likewise, seeds of another specie namely *Z. Myriacanthum* are burnt and its smoke is inhaled for healing syphilitic ulcer of nose. The follicles of *Z. Ailanthoides* SIEB and ZUCC are considered to be pungent and utilized as pepper. Fruits of this plant are ingested to enhance digestion and treat dysentery, diarrhea, and sunstroke. For the treatment of flux and chills, local people living in Asia have used the leaves infused drink of this fruit. Similarly, *Zanthoxylum bungei* PLANCH is believed to possess stimulant and carminative properties and hence improves sweating and is also used as an anthelmintic. Their carpels are infused in vinegar and instilled in ears for the removal of worms and insects. Further, *Z. piperitum* (L.) DC dried follicles were administrated for stimulating urination, dysentery, malaria, and spermatorrhea. For the removal of parasites, carpels of this plant have also been used eternally. Likewise, seeds are also prescribed for the promotion of urination and treatment of dropsy, asthama, and freshens kidney and bladder. In Japan and China, the seeds of *Z. Avicennae* (Lamk.) DC were used as an alternative for pepper.[53]

In China, the history of *Z. bungeanum* cultivation draws back over 2000 years owing to its economical and medicinal values. First, the description of *Z. bungeanum* could be tracked back to Shijing, signifying that this plant is always valued as a remarkable plant owing to its red fruits. In the pre-Qin Dynasty era, folk people most commonly consumed pericarps of this plant as a spice hence revealing its early utilization.[28] Additionally, in Chinese medicine, pericarps are used in medications usually by processing in vinegar and salt water.[67] In traditional Chinese medicine, this plant was used as an herbal medicine having pharmacological properties that improved eyesight, strengthened teeth, and removed cold-dampness.[72] Similarly, *Mingyi Bielu* monograph has reported excellent blood-promotion potential, joint-smoothening activities, and teeth-strengthening activities. Another predominant monograph in Chinese medicine, that is, *Zhenglei Bencao*, revealed the utilization of *Z. bungeanum* for treating vomiting, throat impairment, and abdominal ache. Further, in *Bencao Gangmu*, the pericarps of *Z. bungeanum* have been characterized as a potential treatment against dampness, diarrhea, toothache, and ascariasis.[64] In Western folk medicine, different species of *Zanthoxylum* are used as folk medicine, and hence are classified as "toothache trees" owing to its anesthetic activities that rendered it helpful in minimizing chronic and acute pains.[7]

Pastes prepared from *Zanthoxylum* species are used in Kenya and South Africa for suppressing pains and help in healing wounds,[34] whereas

Nigerians use them to treat anemia, urinary tract infections, toothache, and rheumatism.[2,3,23] Likewise, in Uganda, root-bark of *Zanthoxylum zanthoxyloides* has been used for treating malaria, erectile dysfunctionality, abdominal pain, toothache, dysmenorrhea, and elephantiasis.[4,60] In Cote D'Ivoire, decocted stem of this plant has been reported to reduce toothache and minimize infections resulting from oral pathogens.[38] Additionally, its leaves are prescribed in Togo for treating wounds; however bark of this plant helps in relieving pain.[71] Further, the inhabitants in Ghana used the decocted stem-bark of *Zanthoxylum zanthoxyloides* to treat malaria.[5]

Traditionally, various parts of *Zanthoxylum clava-herculis* are consumed by people of Central African Republic for treating hypertension, diabetes, malaria, and respiratory-related diseases.[39] *Zanthoxylum gilletii* have been used in Côte D'Ivoire for the treatment of malaria, high blood pressure, and skin infections.[74,75] In some regions of Africa, various decocted parts of *Zanthoxylum gilletii* are alone and/or in combination used for treating erectile dysfunctionality.[63] As reported by Kipkore et al.,[36] seeds and barks of *Zanthoxylum chalybeum* were boiled and utilized for treating rheumatism and malaria. Likewise, In South Africa, traditionally, the roots and barks of *Zanthoxylum capense* are used for the treatment of tuberculosis.[12] People of Thailand utilized *Zanthoxylum rhetsa* as condiment and a special spice during cooking as well as for the treatment of different infections.[62] Generally, literature reviewed reveals that folk people more often use various parts of *Zanthoxylum* species as food and medicine. The traditional benefits of some species of genus *Zanthoxylum* are listed in Table 2.1.

TABLE 2.1 Traditional Uses of Different Parts of Some *Zanthoxylum* Species.

Specie name (s)	Area name	Plant part	Traditional uses	References
Zanthoxylum zanthoxyloides	Ghana, Nigeria, Cameroon, Uganda, Togo, Cote D' Ivoire	Stem Leaves Seeds Roots	To treat toothache, malaria, wounds, arthritis, fever, hypertension, and malaria	[2–5,23]
Zanthoxylum gilletii	Kenya, Madagascar, Côte d'Ivoire	Leaves Bark Stem Roots	To treat malaria and fungal infections	[22,47,61,74]

TABLE 2.1 *(Continued)*

Specie name (s)	Area name	Plant part	Traditional uses	References
Zanthoxylum nitidum	China, Thailand, Indonesia, Portugal	Seed Leaves Roots	To treat toothache, ulcer, respiratory diseases, and oral pathogens	[10,27,58]
Zanthoxylum davyi	South Africa	Leaves Roots Bark Seeds	To treat snakebite wounds, coughs, ulcers, toothache, and colds	[63]
Zanthoxylum armatum	Nepal, Abbottabad	Roots Fruits Leaves Bark	To treat diabetes, coughs, and microbial infections	[11,25,50]
Zanthoxylum capense	South Africa	Leaves Bark Roots	To treat asthma, tuberculosis, wounds, and abdominal pains	[1,12,54,57]
Zanthoxylum bungeanum	Japan, China	Fruit Leaves Seeds Bark	To treat fungal and bacterial infections	[29,35,42]
Zanthoxylum tessmannii	Madagascar and Cameroon	Stem Bark	To treat high blood pressure, gonorrhea, erectile dysfunctionality, and inflammations	[21,44,73]
Zanthoxylum buesgenii	South-West Cameroon and Sierra Leone	Leaves Seeds Roots	Enhances fertility in men, treats rheumatism and arthritis	[8,20,55]
Zanthoxylum ovalifolium	India and Australia	Stem Bark Fruit	Helps in treating fungal and bacterial infections	[26,49]
Zanthoxylum avicennae	China	Leaves Bark	Used for treating fungal infections in plants and human beings	[70]
Zanthoxylum usambarense	Kenya	Leaves Bark Fruit Seeds Roots	Utilized to treat malaria	[47]

TABLE 2.1 *(Continued)*

Specie name (s)	Area name	Plant part	Traditional uses	References
Zanthoxylum schinifolium	Asia	Leaves Seeds Roots	To treat jaundice, stomachaches, and diarrhea	[14]
Zanthoxylum clava-herculis	Cameroon	Leaves Stem Bark	Helps in reducing bacterial infections, high blood pressure, and diabetes	[39]

2.4 TRADITIONAL USES OF GENUS MICROMELUM

In Philippines, leaves of Micromelum minutum were consumed to get relief from headaches. In Vietnam, Laos, and Cambodia, leaves of this plant were utilized for easing in itchiness and promotion of menses. However, the people living in Malaysia used to boil roots of this plant to formulate a medicinal paste that was applied for the treatment of ague. Micromelum species are traditionally prescribed for the treatment of fever, ague, ringworm, and for regulating menses.[9,13] It has been reported that *Micromelum minutum* have been used for their medicinal properties in Thai traditional medicine.[30]

In folk medicine, various parts of Micromelum minutum have been used for treating different diseases. Traditionally, its stems were used as a carminative agent while its fruits and flowers acted as a purgative and expectorant, respectively. Locally, leaves' juice was prescribed for the treatment of white scum on tongue, toothache, hemorrhoids, and bad-breath. Generally, leaves are considered to be tonic and shoots were given to treat infantile convulsions. Crushed leaves of Micromelum species were used as an active ingredient of poultice and were consumed to treat skin irritations. For the reduction of irritations resulting due to scabies, leaves of Micromelum species are rubbed on the skin. Leaves and inner bark of twigs were utilized for treating stomachaches, headaches, gonorrhea, thrush, coughs, and colds. Fluids extracted from the bark of Micromelum species were consumed traditionally for the treatment of headaches, while infused bark was used for curing stomachache. Decocted and infused roots of this plant were used to treat diarrhea, and acted as a carminative agent in children. They are known to have medicinal perspectives since ancient times owing to their soothing effect against stomachache, toothache, and headaches. Betel along with roots pieces of this plant have been chewed for the treatment of coughs.[19]

2.5 TRADITIONAL USES OF GENUS CITRUS

Various species of Genus Citrus are known since ancient times owing to their medicinal properties. Among them *Citrus maxima* (Burm.) Merr, *Citrus aurantifolia* (Chirst.) SW., *Citrus limon* (L.) Burm, *Citrus hystrix* DC, and *Citrus mitis* BLCO are the predominant ones in various folk medicines. Fruits of Citrus *maxima* (Burm.) Merr. are characterized as cardio-tonic and are used in treating cough, influenza, and asthma. Rind of *Citrus maxima* (Burm.) Merr. is anthelmintic, prevents vomit, and reduces diarrheal conditions. However, leaves of this fruit were used to treat convulsive cough and epilepsy. Similarly, fruit of another specie of genus citrus, that is, *Citrus aurantifolia* (Chirst.) SW. helped in treating nausea and skin-related issues, while juice of this fruit is considered to be an anthelmintic and antiseptic and therefore used in biliousness and prevents vomiting. Salted peels of this fruit aid digestive system. Further, fruit juice of *Citrus maxima* (Burm.) Merr. along with honey and warm water prevents from catarrhal fever. In some regions of India, leaves of *Citrus maxima* (Burm.) Merr. are utilized for treating infectious disease filariasis. Fruit juice of *Citrus limon* (L.) Burm.f is considered to be an abundant source of Vitamin C that helps in preventing infections, hence making it an effectual tool in treating flu and cold. It has been used for preventing stomach infections, arteriosclerosis, scurvy, and circulation problems. It has also served as antioxidant, antibacterial, and antiseptic agent in ancient times. Fruit juice of *Citrus limon* (L.) Burm.f was used as an appetizer, and for treating vomiting and nausea. In Indonesia, fruit juice of this fruit was used as a tonic ingredient while Thailand people use it as condiment in tom yum.[53]

2.6 TRADITIONAL USES OF GENUS ATALANTIA

Atalantia monophylla DC is a well-known specie among different other species of genus *Atalantia*. Habitats of Malaysia, Vietnam, and Cambodia use the leaves of *Atalantia monophylla* DC for the treatment of lungs related disorders mainly owing to their essential oils that irritate the mucosa and stimulate the motion of bronchial villosities. In Tamil, this plant is traditionally named as kattu elumichai and is conventionally used in the treatment of cough, cold, dyspepsia, arthritis, fever, and rheumatoid. Its fruit juice utilization on daily basis in the evening helps in easing abdominal pains, headaches, itching, nausea, bloating, paralysis, vomits, stomachache, malaria, hemiplegia, joint pains, and skin-related diseases. In ancient scriptures, use of different parts of this plant in treating different bacterial and microbial infections has been reported.[66]

A. monophylla (L) belongs to family Rutaceae and is known as a thorny shrub that is abundantly available in peninsular India.[24] Usage of different parts of *A. monophylla* such as roots, seeds, fruit, leaves, bark, and flowers has been found in several ancient scriptures owing to their medicinal properties. Evidently, oil extracted from *A. monophylla* seeds has been used to treat arthritis.[40] Oil extracted from leaves was used for suppressing the bacterial and fungal growth and reduced paralysis, glandular inflammation, and roots were used as antispasmodic agent.[37,51,56,59]

2.7 TRADITIONAL USES OF GENUS LUNASIA

One of the most renowned species in genus *Lunasia* that is known since ancient times is *Lunasia amara* Blanco. It is a plant species that have been prescribed by traditional practitioners and have been used by people of Manobo and Agusan del Sur for the treatment of different ailments. Understanding regarding traditional knowledge of folk medicine of this plant may be helpful for future drug development.[16]

Indonesians used boiled bark and leaves of *Lunasia amara* in water for the preparation of indigenous lotion that was applied to reduce inflammatory responses. Likewise, Philippines also applied the crushed bark to mitigate the effects of snakebite poison and when ingested prevented stomachaches.[53] Bark of this tree has been used for treating ulcer, snakebites, heartburn, fever, skin ailments, wounds, nausea, gastroenteritis, diarrhea, itchiness, food poisoning, stomachaches, malaria, tetanus, rabies, headaches, and insect venom. Ancient practitioners prepared infusions of tree bark with coconut oil and were tinctured with Kulafu (local wine), as this preparation was affordable. This prepared infusion was applied on affected areas or by ingesting the preparation in case of internal problem.[16]

2.8 TRADITIONAL USES OF GENUS CLAUSENA

People residing in Burma use *Clausena excavate* for the treatment of stomachaches and stomach-related ailments. In countries like Vietnam and Cambodia, *Clausena excavate* were utilized as a stimulant and was used for easiness during menses and digestion and for treating paralysis. Likewise, Taiwan residents used the decocted roots of this plant in formulating a drink that helped in promoting sweating. Malays used its roots and leaves for healing of sores, mitigation of headaches, and strengthening post-delivery conditions. However, Indonesians used the extracted juice from leaves of this plant in the reduction of fever and to decrease intestinal worms. Fruits of another specie of genus *Clausena*, that is, *Clausena lansium* were ingested by Chinese as an antidote when consumed in empty stomach. Similarly, in the Pentsao, fruits of *Clausena lansium* were believed to possess anthelmintic potential and fruits are mentioned as a stomachic, cooling, and anthelmintic remedy.[69]

2.9 TRADITIONAL USES OF MISCELLANEOUS GENUS

Some other species and genus in family Rutaceae that have predominant position in different traditional medications like Unani, Ayruvadic, and Chinese are *Limonia acidissima, Murraya paniculata, Euodia elleryana,* and *Paramignya*. Among these, unripe *Limonia acidissima* fruits were administrated to treat dysentery and diarrhea. But seeds and leaves of this plant were consumed for treating heart ailments, vomits, hiccups, and indigestions. Pulp of its fruit were applied topically to minimize the venomous effects of snake, insect, and reptile bites and have also been used for treating biliousness. Furthermore, pulp have been used to treat cold, coughs, asthma, diarrhea, inflammations, and leucorrhea.[53] Similarly, leaves of *Murraya paniculata*, another specie in family Rutaceae are prescribed as a stimulant and effectively treated toothache, dysentery, and diarrhea. Decocted leaves were taken as dropsy while powdered leaves were applied to fresh-cuts. Grinded roots were applied topically for treating body aches. However, flowers and bark are used in cosmetic products.[53]

Paramignya species belong to family Rutaceae and are found throughout South Vietnam, Malaysia, Australia, Thailand, Philippines, and Sri Lanka. Genus *Paramignya* has well-established medicinal values in different countries around the world. Local practitioners prescribed *P. trimera roots* as a traditional medicine for the treatment of diabetes and hepatic disorders. Likewise, people of Thailand utilized *P. grithii* stems to treat nose infections.[15,33,68]

KEYWORDS

- **Rutaceae**
- **Ayurvedic**
- **Unani**
- **traditional Chinese usages**
- **malaria**
- **skin ailments**

REFERENCES

1. Adebayo, S. A.; Dzoyem, J. P.; Shai, L. J.; Eloff, J. N. The Anti-inflammatory and Antioxidant Activity of 25 Plant Species used Traditionally to Treat Pain in Southern African. *BMC Complement. Altern. Med.* **2015,** *15*, 159.
2. Adesina, S. K. The Nigerian Zanthoxylum: Chemical and Biological Values. *Afr. J. Tradit. Complement. Altern. Med.* **2005,** *2,* 282–301.
3. Ameh, S. J.; Tarfa, F. D.; Ebeshi, B. U. Traditional Herbal Management of Sickle Cell Anemia: Lessons from Nigeria. *Anemia* **2012,** *2012*, 607436.
4. Andima, M.; Coghi, P.; Yang, L. J.; Wong, V. K. W.; Ngule, C. M.; Heydenreich, M.; Ndakala, A. J.; Yenesew, A.; Derese, S. Antiproliferative Activity of Secondary Metabolites from Zanthoxylum zanthoxyloides Lam: In Vitro and In Silico Studies. *Phcog. Commun.* **2020,** *10*, 44–51.
5. Asase, A.; Oppong-Mensah, G. Traditional Antimalarial Phytotherapy Remedies in Herbal Markets in Southern Ghana. *J. Ethnopharmacol.* **2009,** *126*, 492–499.
6. Baliga, M. S.; Bhat, H. P.; Joseph, N.; Fazal, F. Phytochemistry and Medicinal uses of the Bael Fruit (Aegle marmelos Correa): A Concise Review. *Food Res. Int.* **2011,** *44* (7), 1768–1775.
7. Bautista, D. M.; Sigal, Y. M.; Milstein, A. D.; Garrison, J. L.; Zorn, J. A.; Tsuruda, P. R.; Nicoll, R. A.; Julius, D. Pungent Agents from Szechuan Peppers Excite Sensory Neurons by Inhibiting Two-Pore Potassium Channels. *Nat. Neurosci.* **2008,** *11*, 772–779.

8. Burkill, H. M. In *The Useful Plants of West Tropical Africa, Families M-R,* 2nd ed.; The Royal Botanic Garden: Kew, UK, 1998; vol 4, pp 1–969.
9. Cambie, R. C.; Ash J. In *Fijian Medicinal Plants*; CSIRO Publishing, 1994; pp 2031–2032.
10. Chakthong, S.; Ampaprom, S.; Inparn, S.; Phetkul, U.; Chusri, S.; Limsuwan, S.; Voravuthikunchai, S. P. New Alkylamide from the Stems of Zanthoxylum nitidum. *Nat. Prod. Res.* **2019,** *33*, 153–161.
11. Choi, H. J.; Song, J. H.; Kwon, D. H.; Baek, S. H.; Ahn, Y. J. Antiviral Activity of Zanthoxylum Species against Influenza vVirus. *Korean J. Med. Crop Sci.* 2008, *16*, 273–278.
12. Cock, I. E.; Van Vuuren, S. F. The Traditional use of Southern African Medicinal Plants for the Treatment of Bacterial Respiratory Diseases: A Review of the Ethnobotany and Scientific Evaluations. *J. Ethnopharmacol.* **2020,** *263*, 113204.
13. Croft, K. D.; Toia, R. F. Coumarins from Micromelum minutum. *Planta Medica* **1989,** *55*, 401.
14. Cui, H. Z.; Choi, H. R.; Choi, D. H.; Cho, K. W.; Kang, D. G.; Lee, H. S. Aqueous Extract of Zanthoxylum schinifolium Elicits Contractile and Secretory Responses via _1-adrenoceptor Activation in Beating Rabbit Atria. *J. Ethnopharmacol.* **2009,** *126*, 300–307.
15. Dang, P. H.; Le, T. H.; Phan, K. P. T. Two Acridones and Two Coumarins from the Roots of Paramignya trimera. *Tetrahedron Lett.* **2017,** *58,* 1553–1557.
16. Dapar, M. L. G.; Demayo, C. G. Folk Medical uses of Lunas Lunasia amara Blanco by the Manobo People, Traditional Healers and Residents of Agusan del Sur, Philippines. *Sci. Int.* **2017,** *29* (4), 823–826.
17. Das, B.; Das, R. Medicinal Properties and Chemical Constituents of Aegle marmelos Correa. *Indian Drugs* **1995,** *32*, 93–99.
18. Dhankhar, S.; Ruhil S.; Balhara, M.; Dhankhar, S.; Chhillar, A. Aegle marmelos (Linn.) Correa: A Potential Source of Phytomedicine. *J. Med. Plant Res.* **2011,** *5* (9), 1497–1507.
19. Fatema-Tuz-Zohora, H. C.; Ahsan, M. Chemical Constituents, Cytotoxic Activities and Traditional uses of Micromelum minutum (Rutaceae): A Review. *Pharm. Pharmacol. Int. J.* **2019,** *7* (5), 229–236.
20. Fonge, B. A.; Egbe, E. A.; Fongod, A. G. N.; Focho, D. A.; Tchetcha, D. J.; Nkembi, L.; Tacham, W. N. Ethnobotany Survey and uses of Plants in the Lewoh-Lebang Communities in the Lebialem Highlands, SouthWest Region, Cameroon. *J. Med. Plant Res.* **2012,** *6*, 855–865.
21. Fouda, Y. B.; Tom, E. N. L.; Oumarou, B. A.; Mbolang, L. N.; Kuissu, A. M. T.; Djomeni, P. D. D.; Dimo, T. Aqueous Extract of Fagara tessmannii Engl. (Rutaceae) Exhibits Antihypertensive Activity in NO Synthase Inhibitor-Induced Hypertensive Rats. *J. Integr. Cardiol.* **2020,** *3*, 2–9.
22. Gaya, C.; Kawaka, J.; Muchugi, A.; Ngeranwa, J. Variation of Alkaloids in the Kenyan Zanthoxylum gilletii (De wildWaterman). *Afr. J. Plant Sci.* **2013,** *7*, 438–444.
23. Gbadamosi, I. T. An Inventory of Ethnobotanicals used in the Management of Sickle Cell Disease in Oyo State, Nigeria. *Bot. Res. Int.* **2015,** *8*, 65–72.
24. Ghosh, A. Survey of Ethno-Medicinal Climbing Plants in Andaman Nicobar Islands, India. *Int. J. Pharm. Life Sci.* **2011,** *17*, 621–630.

25. Guleria, S.; Tiku, A. K.; Kurl, A.; Gupta, S.; Sign, G.; Razdan, V. K. Antioxidant and Antimicrobial Properties of the Essential Oil and Extracts of Zanthoxylum alatum Grown in North-Western Himalaya. *Sci. World J.* **2013,** *2013*, 790580.
26. Halstead, C. W.; Forster, P. I.; Waterman, P. G. Alkaloids from the Stem Bark of an Australian Population of Zanthoxylum ovalifolium. *Nat. Prod. Res.* **2006,** *20*, 940–945.
27. Han, Z. Z.; Li, R. L.; Yang, T. C.; Zhan, R. T.; Chen, W. W. The Effect of Zanthoxylum Nitidum on Gastric Ulcer Rats Induced by Hydrochloric Acid and Ethanol. *J. Guangzhou Univ. Tradit. Chinese Med.* **2012,** *29*, 292–294.
28. Huang, D. M.; Zhao, G. H.; Chen, Z. D.; Kan, J. Q. The Food Culture of Zanthoxylum bungeanum in China. *China Condiment* **2016,** *1*, 75–81.
29. Hwang, K. A.; Kwon, J. E.; Noh, Y.; Park, B.; Jeong, Y. J.; Lee, S. M.; Kim, S. Y.; Kim, I.; Kang, S. C. Effects of Zanthoxylum piperitum Ethanol Extract on Osteoarthritis Inflammation and Pain. *Biomed. Pharmacol. J.* **2008,** *105*, 481–490.
30. Itharat, A.; Houghton, P. J.; Eno-Amooquaye, E. In Vitro Cytotoxic Activity of Thai Medicinal Plants used Traditionally to Treat Cancer. *J. Ethnopharmacol.* **2004,** *90* (1), 33–38.
31. Kar, A.; Choudhry, B. K.; Bandhopadhyay, N. G. Comparative Evaluation of Hypoglycemic Activity of Some Indian Medicinal Plants in Alloxan Diabetic Rats. *J. Ethnopharmacol.* **2003,** *84*, 105–108.
32. Karunanayake, E. H.; Welihinda, J.; Sirimanne, S. R.; Sinnadorai, G. Oral Hypoglycemic Activity of Some Medicinal Plants of Sri Lanka. *J. Ethnopharmacol.* **1984,** *11*, 223–231
33. Khoi, N. M.; Hang, P. T. N.; Phuong, D. T. Study on Acute Toxicity, Hepatoprotective Activity and Cytotoxic Activity of Paramignya trimera. *J. Med. Mater.-Hanoi* **2013,** *18*, 14–20.
34. Kigen, G.; Kipkore, W.; Wanjohi, B.; Haruki, B.; Kemboi, J. Medicinal Plants used by Traditional Healers in Sangurur, Elgeyo Marakwet County, Kenya. *Pharm. Res.* **2017,** *9*, 333–347.
35. Kim, M. H.; Lee, H. J.; Park, J. C.; Hong, J.; Yang, W. M. Zanthoxylum Piperitum Reversed Alveolar Bone Loss of Periodontitis via Regulation of Bone Remodeling-Related Factors. *J. Ethnopharmacol.* **2017,** *195*, 137–142.
36. Kipkore, W.; Wanjohi, B.; Rono, H.; Kigen, G. A Study of the Medicinal Plants used by the Marakwet Community in Kenya. *J. Ethnobiol. Ethnomed.* **2014,** *10*, 24.
37. Kirtikar, K. R.; Basu, B. D. In *Indian Medicinal Plants*; Bishen Sing Mahendra Pal Sing Publication: Dehradun, 2010; vol 1, pp 1655–1656.
38. Kiyinlma, C.; Chimène, A.; Allassane, D.; Justin, K. N. Ethnobotanical Study and Valorization of Medicinal Plants from the Classified Forest of Foumbo (Northern Cote D'ivoire). *Int. J. Rec. Sci. Res.* **2020,** *11*, 38848–38853.
39. Kosh-Komba, E.; Touckia, I.; Worowounga, X.; Toumnou, A. L.; Mololi, A.; Mukeina, G.; Semballa, S.; Batawilla, K.; Akpagana, K. Case Study and Phytochemical Investigation of Zanthoxylum zanthoxyloids and Zanthoxylum macrophylum (Rutaceae) in the Central African Republic and Togo: A Comparative Approach. *Am. J. Phytomed. Clin. Ther.* **2017,** *6*, 1.
40. Kumar, R. B.; Narayana, B. S. Tribal Medicinal Studies on Sriharikota is Land, Andhra Pradesh. *Ethnobot. Leafl.* **2010,** *14*, 95–107.

41. Lampronti, I.; Martello, D.; Bianchi, N.; Borgatti, M.; Lambrtini, E.; Piva, R.; Jabbars, S.; Choudhuri, M. S.; Khan, M. T.; Gambari, R. In Vitro Antiproliferative Effect on Human Tumor Cell Lines of Extracts from the Bangladesi Medicinal Plant Aegle marmelos Correa. *Phytomedicine* **2003,** *10,* 300–308.
42. Lee, S.; Lim, K. Glycoprotein of Zanthoxylum piperitum DC has a Hepatoprotective Effect Via Anti-Oxidative Character In Vivo and In Vitro. *Toxicology In Vitro* **2008,** *22*, 376–385.
43. Maity, P.; Hansda, D.; Bandyopadhyay, U., Mishra, D. K. Biological activities of crude extracts and chemical constituents of Bael Aegle marmelos (L.) Corr. *Indian J. Exp. Biol.* **2009,** *47*, 849–861.
44. Massoma, L.; Gasco, M.; Rubio, J.; Yucra, S.; Sock, E. N.; Gonzales, G. F. Effect of the Ethanolic Extract from Fagara Tessmannii on Testicular Function, Sex Reproductive Organs and Hormonal Level in Adult Male Rats. *J. Androl.* **2011,** *43*, 139–144.
45. Nanjing University of Traditional Chinese Medicine. Chinese Medicine Dictionary, Science and Technology Press of Shanghai: Shanghai, China, 2006; p 1469.
46. Okagu, I. U.; Ndefo, J. C.; Aham, E. C.; Udenigwe, C. C. Zanthoxylum Species: A Comprehensive Review of Traditional Uses, Phytochemistry, Pharmacological and Nutraceutical Applications. *Molecules* **2021,** *26* (13), 4023.
47. Omara, T. Antimalarial Plants used Across Kenyan Communities. *Evid. Based Complementary Altern. Med.* **2020,** *2020*, 4538602.
48. Parichha, S. Bael (Aegle Marmelos): Nature's Most Natural Medicinal Fruit. *Orissa Review*, 2004, vol 9, pp 16–17.
49. Pavani, P.; Naika, R. Evalution of Anti-bacterial Activity of zanthoxylum ovalifolium wight (Rutaceae) Against Selected Pathogenic Bacteria. *Plant Arch.* **2020,** *20*, 2591–2594.
50. Phuyal, N.; Jha, P. K.; Raturi, P. P.; Rajbhandary, S. Total Phenolic, Flavonoid Contents, and Antioxidant Activities of Fruit, Seed, and Bark Extracts of Zanthoxylum armatum DC. *Sci. World J.* **2020,** *2020*, 8780704.
51. Prasad, Y. R. Chemical Investigation and Antimicrobial Efficacy of the Volatile Leaf Oil of Atalantia monophylla Corr. *Parfümerie und Kosmetik* **1988,** *69*, 418–419.
52. Rahman, S.; Parvin, R. Therapeutic Potential of Aegle marmelos (L.)—An overview. *Asian Pac. J. Trop. Dis.* **2014,** *4* (1), 71–77.
53. Roy, D.; Rahman, A. H. M. M. Systematic Study and Medicinal uses of Rutaceae Family of Rajshahi District, Bangladesh. *Plant Environ. Dev.* **2016,** *5* (1), 26–32.
54. Sagbo, I. J.; Mbeng, W. O. Plants used for Cosmetics in the Eastern Cape Province of South Africa: A Case Study of Skin Care. *Phcog. Rev.* **2018,** *12*, 139–156.
55. Sandjo, L. P.; Kuete, V.; Tchangna, R. S.; Efferth, T.; Ngadjui, B. T. Cytotoxic Benzophenanthridine and Furoquinoline Alkaloids from Zanthoxylum buesgenii (Rutaceae). *Chem. Cent. J.* **2014,** *8*, 61.
56. Sankaranarayanan, S. P.; Bama, J.; Ramachandran, P. T.; Kalaichelvan, M.; Deccaraman, M.; Vijayalakshimi, R.; Dhamotharan, B.; Dananjeyan, S.; Bama, S. Ethnobotanical Study of Medicinal Plants used by Traditional users in Villupuram District of Tamil Nadu. *J. Med. Plants Res.* **2010,** *4,* 1089–1101.

57. Semenya, S. S.; Maroyi, A. Plants used by Bapedi Traditional Healers to Treat Asthma and Related Symptoms in Limpopo Province, South Africa. *Evid. Based Complementary Altern. Med.* **2018,** *2018*, 2183705.
58. Sepsamli, L.; Prihastanti, J. E. Ethnobotany of Balimo (Zanthoxylum nitidum) in the Kanayatn Dayak Community in Tapakng, West Kalimantan. *Biosaintifika: J. Biol. Biol. Educat.* **2019,** *11*, 318–324.
59. Sukumaran, S.; Raj, A. D. Medicinal Plant of Sacred Groves in Kanyakumari District South Western Ghats. *Indian J. Tradit. Knowl.* **2008,** *9*, 294–299.
60. Supabphol, R.; Tangjitjareonkun, J. Chemical Constituents and Biological Activities of Zanthoxylum limonella (Rutaceae): A Review. *Trop. J. Pharm. Res.* **2014,** *13*, 2119–2130.
61. Tankeo, S. B.; Damen, F.; Awouafack, M. D.; Mpetga, J.; Tane, P.; Eloff, J. N.; Kuete, V. Antibacterial Activities of the Methanol Extracts, Fractions and Compounds from Fagara tessmannii. *J. Ethnopharmacol.* **2015,** *169*, 275–279.
62. Tantapakul, C.; Phakhodee, W.; Ritthiwigrom, T.; Yossathera, K.; Deachathai, S.; Laphookhieo, S. Antibacterial Compounds from Zanthoxylum rhetsa. *Arch. Pharm. Res.* **2012,** *35*, 1139–1142.
63. Tarus, P. K.; Coombes, P. H.; Crouch, N. R.; Mulholland, D. A. Benzo[c] phenanthridine Alkaloids from Stem Bark of the Forest Knobwood, Zanthoxylum davyi (Rutaceae). *South Afr. J. Bot.* **2006,** *72*, 555–558.
64. Tong, X. R.; Wang, P. M. A Preliminary Discussion on the Wentongsanjie Effect of Zanthoxylum bungeanum. *Lishizhen Med. Materia Medica Res.* **1999,** *12*, 897–898.
65. Verma, K. K.; Kumar, B.; Raj, H.; Sharma, A. A Review on Chemical Constituents, Traditional Uses, Pharmacological Studies of Zanthoxylum armatum (Rutaceae). *J. Drug Deliv. Ther.* **2021,** *11* (2-S), 136–142.
66. Vijayakumar, S.; Arulmozhi, P.; Rajalakshmi, S.; Mahadevan, S.; Parameswari, N. A Review of the Taxonomy, Ethno-botany and Pharmacological Activity of Atalantia monophylla L. *Acta Ecologica Sinica* **2020,** *40* (3), 204–209.
67. Wang, X. T. Processed Methods of Traditional Chinese Medicine; Science and Technology Press of Jiangxi: Nanchang, China, 1989; pp 227–228.
68. Wattanapiromsakul, C.; Waterman, P. G. Flavanone, Triterpene and Chromene Derivatives from the Stems of Paramignya grithii. *Phytochemistry* **2000,** *55*, 269–273.
69. Wiart, C. In *Ethnopharmacology of Medicinal Plants: Asia and the Pacific*; Springer Science & Business Media, 2007.
70. Xiong, Y.; Huang, G.; Yao, Z.; Zhao, C.; Zhu, X.; Wu, Q.; Zhou, X.; Li, J. Screening effective antifungal substances from the bark and leaves of Zanthoxylum avicennae by the bioactivity-guided isolation method. *Molecules* **2019,** *24*, 4207.
71. Yaovi, N. Characteristics, Fatty Acid Profile, Strategic Importance of Zanthoxylum zanthoxyloides (rutaceae) Seed Oil and Sustainable Conservation of the Species. *Int. J. Dev. Res.* **2018,** *8*, 21425–21429.
72. Zhonghua bencao Commission. In *Chinese Materia Medica*; Science and Technology Press of Shanghai: Shanghai, China, 1999; p 976.
73. Zhou, X. J.; Chen, X. L.; Li, X. S.; Su, J.; He, J. B.; Wang, Y. H.; Li, Y.; Cheng, Y. X. Two Dimeric Lignans with an Unusual _,_-Unsaturated Ketone Motif from Zanthoxylum podocarpum and their Inhibitory Effects on Nitric Oxide Production. *Bioorg. Med. Chem. Lett.* **2011,** *21*, 373–376.

74. Zirihi, G. N.; N'guessan, K.; Etien, D. T.; Serikouassi, B. P. H. Evaluation In Vitro of Antiplasmodial Activity of Ethanolic Extracts of Funtumia elastica, Rauvolfia vomitoria and Zanthoxylum gilletii on Plasmodium Falciparum Isolates from Côted'Ivoire. *J. Anim. Plant Sci.* **2009,** *5*, 406–413.
75. Zirihi, G.; Yao, D.; Kra-adou, K.; Grellier, P. Phytochemical and Pharmacological Studies of Alcoholic Extract of Fagara macrophylla (Oliv) Engl (Rutaceae): Chemical Structure of Active Compound Inducing Antipaludic Activity. *J. Chinese Clin. Med.* **2007,** *2*, 205–210.

CHAPTER 3

Ethnomedicinal Applications of the Family Rutaceae

MUHAMMAD RIZWAN[1], TAJ UDDIN[1], ZAFAR ALI[1], MOHAMMAD ALI[1], FAZAL AKBAR[1], ABDUR RAUF[2], and MOHAMMED A. AL-DUAIS[3,4]

[1]*Centre for Biotechnology and Microbiology, University of Swat, Khyber Pakhtunkhwa, Pakistan*

[2]*Department of Chemistry, University of Swabi, Anbar, Swabi, Khyber Pakhtunkhwa, Pakistan*

[3]*Department of Biochemistry, Faculty of Science, University of Tabuk, Tabuk, Saudi Arabia*

[4]*Biochemistry Unit, Chemistry Department, Faculty of Science, Ibb University, Ibb, Yemen*

ABSTRACT

The use of herbal plants for therapeutic purposes has been documented in many ethnomedical studies around the world. Traditional knowledge tends to be transmitted verbally and is often innate. A modern ethnomedical approach allows for the development of innovative drugs and treatments as well as preserves medicinal plants by recording how plants have been used for centuries. The members of family Rutaceae have been used in traditional medicines for prevention and treatment of various diseases like diarrhea, abdominal pain, stomachache, etc. This chapter summarizes the research documents on ethnomedicinal importance of the Rutaceae family.

Phytochemical and Pharmacological Investigation of the Family Rutaceae.
Abdur Rauf, Hafiz Ansar Rasul Suleria, & Syed Muhammad Mukarram Shah (Eds.)

More detailed research is needed to assess the real therapeutic potential of the Rutaceae family.

3.1 INTRODUCTION

One of the factors influencing human civilization is the interaction between humans and plants, particularly in the medical field.[28] The use of herbal plants for therapeutic purposes has been documented in many ethnomedical studies around the world, which has been used in their respective cultures for generations.[49] This traditional knowledge is extensive in Pakistan, but it is difficult to access because of the diversity of cultures.[10] Furthermore, traditional knowledge tends to be transmitted verbally and is often innate.[20] Due to this, tribal leaders, village heads, elders, heads of kampong (small villages), or traditional healers within each particular community often possess and own the knowledge.[23]

In traditional medicine, the plant's parts (stem, leaves, flowers, fruits, roots, and barks) are used to treat a wide range of illnesses.[9] When treating a specific disease, sometimes two or more parts of a plant are mixed into powder or liquid form and given to the patients.[51] A modern ethnomedical approach allows for the development of innovative drugs and treatments as well as preserves medicinal plants by recording how plants have been used for centuries.[59]

This chapter summarizes the ethnomedicinal importance of some genus and species of the Rutaceae family.

3.2 ETHNOMEDICINAL IMPORTANCE OF RUTACEAE

The Rutaceae family is a diverse family with a variety of plants. The majority of these herbs have been used for the treatment of various diseases for a long time. Some genera of the Rutaceae family and their use in traditional medicines are summarized here.

3.2.1 ETHNOMEDICINAL VALUES OF ZANTHOXYLUM GENUS

Zanthoxylum species have been used as folk medicine in Thailand for centuries.[38] There are febrifugal, sudorific, and diuretic properties in the bark, while dental caries can be treated with the essential oil of the fruit. Traditionally, the bark of the tree has been used to treat cardiac, respiratory, and stomach infections, as well as rheumatism and tooth infections.

As well as spice, the essential oil extracted from the fruits is known as "Mullilam oil," which has medicinal benefits including antiseptic, antidiarrheal, anticholera, anti-inflammatory, and soothing agent for dental caries. By rubbing the hard spines against rocks and adding water, the Kanikkar tribe prepares a paste that is applied to a nursing mother's breast for pain relief and to increase milk supply. A mixture of pounded bark and oil is a good remedy for stomach aches in the Philippines. Bark decoctions are also used as antidotes for snake bites and the bark is chewed for chest pain.[5,25,34,40,55,60] Table 3.1 shows the ethnomedicinal importance of species belonging to genus *Zanthoxylum*.

TABLE 3.1 Ethnomedicinal Values of Genus *Zanthoxylum*.

Plant species	Ethnomedicinal values of plant's part
Z. acanthopodium	Roots are used for diabetes, toothache, and stomachache.
Z. ailanthoides	Fruits and barks are used for fever, diarrhea, vomiting, snake bites, abdominal pain, and epigastric pain.
	Stem is used for reduction in heart attacks and improvement in bone injury.
Z. alatum	Seeds are used for stomach problems, cholera, and fever.
	Branches and fruits are used for toothache, anorexia, ataxia, abdominal pain, skin diseases, and stomach problems.
	Roots are used for the treatment of digestive system's diseases and poisonous snake bites.
Z. americanum	The whole plant is used for the treatment of malaria, burns, sore throat, toothaches, and rheumatic conditions.
Z. armatum	Seeds and fruits are used for roundworms, skin diseases, dyspepsia, and as a tonic in fever.
	Seeds, branches, and barks are used in stomach problems, as an anthelmintic and carminative.
	The whole plant is used for fever, dysuria, cough, headache, asthma, microbial infections, diabetes, toothache, cancer, diarrhea, abdominal pain, cholera, hepatosis, and worms, as well as to promote arterial blood flow to affected areas (vasodilation), prevent heart attacks, as analgesics, anti-inflammatory, pesticides, and tonics for the stomach.
Z. avicennae	Stem and branches are used for the treatment of snake bite and as a tonic for stomach.
Z. beecheyanum	Leaves are used for the treatment of skin diseases and bellyache.
Z. budrunga	Leaves are used for the treatment of diarrhea and dyspepsia.
	Bark of the stem is used in headache, cough, and dysentery.

TABLE 3.1 *(Continued)*

Plant species	Ethnomedicinal values of plant's part
Z. bungeanum	Fruits are used for the treatment of abdominal pain, stomach ache, toothache, and vomiting.
Z. capense	Leaves are used for the treatment of epilepsy, toothache, flatulent colic, stomachache, and fever.
Z. caribeum	Stem bark and leaves are used to treat skin ulcers, herpes, fever, spasm, and asthma.
	The wood is also used for diseases of the skin.
Z. chalybeum	Leaves are used for the treatment of pneumonia and severe cold.
	Bark is used for the treatment of dizziness, coughs, cold, malaria, and as an appetizer for children.
	Roots are used for diarrhea of goats.
	Fruits are used for headache, wounds healing, toothache, cough, colds, and malaria.
Z. chiloperone	Roots are used as antirheumatic and antimalarial agents.
Z. davyi	Leaves are used for the treatment of infected wounds, chest pain, colds, severe cough, and snake bites.
	Spines are used for the treatment of toothache and pleurisy.
	Bark of the stem is used for the treatment of sore throat and mouth ulcer.
	Bark of the roots and roots are used in the treatment of toothache.
Z. dugandii	Bark is used as a diuretic and sudorific.
Z. ekmanii	Roots and leaves are used for the treatment of toothache and malaria.
Z. fagara	Seeds, fruits, and leaves are used as sudorific and sedative.
Z. gillettii	Leaves are used for the treatment of hypertension, and also used as an analgesic.
Z. hyemale	Leaves are used as a pain reliever, sudorific, emetic, and salivator.
Z. integrifoliolum	Aboriginals of the Yamei tribes use bark to treat snake bites.
Z. leprieurii Guill	Leaves are used for treating bilharzia, gingivitis, and stomatitis.
	Roots are used as an antiulcer, urinary antiseptic, antibacterial, and antisickling agent.
	Stem bark is used as an antimicrobial, digestion aid, antidiarrheal, anticancer, antiodontologic, and antiparasitic.
Z. liebmanianum	Bark is used to treat parasitic infections such as amebiasis, and also used in local anesthesia.
Z. limonella	Bark is used as an antifungal, diuretic, and sudorific.
Z. macrophylla	Barks and seeds are used for urogenital infections, rheumatism, stomachache, malaria, fever, and toothache.

TABLE 3.1 *(Continued)*

Plant species	Ethnomedicinal values of plant's part
Z. monophyllum	Bark is used to treat ophthalmia, jaundice, and runny nose. Also used as an anesthetic.
Z. myricanthum	Seeds are used as inhaler (smoke) for treating incarnated nose.
Z. naranjillo	Leaves are used to treat inflammations and associated infections.
Z. nitidum	Fruits are used for the treatment of colic, cough, diarrhea, vomiting, and stomachache.
	Roots are used to treat paresis, rheumatism, fever, stomachache, and toothache.
	Stem, seeds, and branches are used in cholera, diarrhea, and fever.
Z. piperitum	All parts of the plant are used in abdominal pain, diarrhea, and vomiting.
Z. rhetsa	Spines are used to give relief from breast pain and increase lactation in nursing mothers.
	Seeds are used for the treatment of rheumatism, toothache, and asthma.
	Fruits are used for the treatment of dyspepsia, some types of diarrhea, UTI, and digestion problems.
	Bark is masticated and applied to snake bites.
Z. riedelianum	The whole plant is used in skin stains, rheumatism, and inflammations.
Z. rigidum	Leaves are used for treating toothache.
Z. rhoifolium	Root bark is used in malaria, inflammations, and as a tonic.
	Bark is used to treat tumor, ear pain, and toothache.
Z. scandens	Leaves, stems, and roots are used to treat traumatic injuries, rheumatism, toothache, and abdominal pain.
Z. schinifolium	Ripen pericarp and leaves are used in epigastric pain.
Z. simulans	Roots are used in gastrointestinal disorders and snake bites.
Z. tessmannii	Stem bark is used for the treatment of inflammations, swelling, and tumor.
Z. tetraspermum	Stem bark is used in fungal infections, hypertension, parasitic infections, diarrhea, and muscular pain. Also used as an antiplatelet, diuretic, sudorific, analgesic, and antispasmodic.
Z. usambarense	Bark is used in the treatment of rheumatism.
	Seeds are used in catarrhal fever, malaria, and respiratory tract infections.
Z. xanthoxyloides	Bark and leaves are used in snake bite, toothache, colds, fever, and cough.
	Leaves are also used as laxative, astringent, antiseptic, and scarring.
	Roots are used as a digestive aid, tooth cleaner, antisickler, and antiseptic.
	Stem bark is used in rheumatism, parasitic infections, and UTI. It is also used as antidiuretic, digestive aid, and antiodontalgic.

3.2.2 ETHNOMEDICINAL VALUES OF CALODENDRUM GENUS

The *Calodendrum* genus has been used medicinally in South Africa, Kenya, and Eswatini.[29] *Calodendrum capense* infusion and/or decoction are commonly used for treating skin infections, toothache, cough, fever, and stomach complaints. Leaf, fruit, seed, and stem bark extracts are also used to treat insect bites. A number of other uses of the species are listed in traditional medicine, such as for emetic and hair softening effects, as well as for easing childbirth, impotency, sterility, and snake bite.[2,8,31,32]

3.2.3 ETHNOMEDICINAL VALUES OF CLAUSENA GENUS

Clausena excavata has been used in traditional medicine to treat abdominal pain, snake bites, and as an aid in detoxing.[54,58] Roots and leaves are used as poultices for sores, including ulcerations of the nose. Traditionally, root decoction is drunk for colic and bowel complaints. Gimlets were discovered in Kelantan as a yawner.[43] A decoction of the leaves and flowers can be used for colic, and after childbirth, a decoction of the flowers can be used. Colds, abdominal pains, malaria, and dysentery are treated with the leaves of this plant.[14] In patients with decayed teeth, powdered rootstock can be used as a treatment, while its stem is given to patients with colic or diarrhea. In Java, coughs are treated with the juice of the plant as a vermifuge.[56,61]

Different African countries traditionally use *Clausena anisata* (Willd.) to treat a variety of diseases. The Nigerian mixture of *C. anisata*, *Afraegle paniculata*, and *Azadirachtha indica* functions as an anti-gut disorder and antimalarial agent. *C. anisata* is a traditional remedy for oral candidiasis and skin infections caused by fungi in Tanzania.[14] Traditional healers use *C. anisata* as an anticonvulsant and antiepileptic drug in Temeke (Daressalam, Tanzania).[42] High blood pressure is treated with *C. anisata* leaves in South Africa.[30]

Clausena lansium has been used traditionally for a variety of purposes, including medicine. Traditionally, the fruit has stomach-soothing and cooling properties, and it was regarded as a vermifuge. According to Chinese folklore, eating wampee can counteract the effects of eating too many lychees. Vietnamese and Chinese medicines make use of half-cut, sun-dried, and immature fruits as bronchitis remedies. To remove dandruff and to keep the hair's color, the decoction of leaves is used as a hair wash.[33,39,47]

Clausena harmandiata is traditionally used as a remedy for food poisoning and digestive gas. Additionally, the roots are effective in curing fever, headaches, and eye pain.[4,16]

The Philippines traditionally use *Clausena anisum-olens* for its medicinal benefits. In addition to being stuffed into pillows, the leaves are also used in baths against rheumatism or in decoctions to treat nausea during pregnancy. A decoction of the fruits and roots is used to treat coughs with fever.[22,63,64]

3.2.4 *CITRUS GENUS*

Tooth powder made from *Citrus limon* promotes oral health and treatment of scurvy.[41] The fruits and leaves of *Citrus medica* help to treat vomiting, tumors, skin and pulmonary diseases, piles, fever, dysmenorrhea, dysentery, constipation, diarrhea, cancer, asthma, worm infestations, and anorexia. *Citrus aurantium* is used to suppress appetite and stimulate mental activity. In traditional Chinese medicine, it is also used to treat nausea, indigestion, constipation, cardiovascular diseases, and cancer.[13,37]

3.3 ETHNOMEDICINAL IMPORTANCE OF OTHER SPECIES OF THE RUTACEAE FAMILY

3.3.1 *AEGLE MARMELOS*

Uses in ethnomedicine: Ripe fruits can improve constipation. It is effective in cleaning and toning the intestines. In order to achieve optimal results, Sherbet made from the pulp of the ripe fruit should be consumed. Chronic diarrhea and dysentery can be treated effectively using the unripe or half-unripe fruit. In traditional medicine, bael leaves are believed to be effective for treating peptic ulcers. Several home remedies use the root of this tree to cure ear problems. Bael leaves are used to prepare a medicated oil that treats respiratory issues and recurrent colds.[12,15,27,48]

FIGURE 3.1 Fruits and leaves of *Aegle marmelos*.

Image credit: https://plantskingdom.in/products/bilva-avenue-trees

3.3.2 *CITRUS MAXIMA*

Fruits are believed to be nutritive, cardiotonic, and useful for treating asthma, cough, and influenza. It is a good remedy for vomiting, diarrhea, and stomach cramps. Epilepsy and convulsive cough can be treated with the leaves.[1,11,52]

FIGURE 3.2 Fruits and leaves of *Citrus maxima*.

Source: Image by Bùi Thụy Đào Nguyên. https://creativecommons.org/licenses/by-sa/3.0/

3.3.3 *CITRUS AURANTIFOLIA*

Fruits are used to relieve sore throats, biliousness, nausea, irritation of the skin, eye complaints, and vomiting. The juice is anthelmintic, antiseptic, and stomachic. The remedy for catarrhal fever is to drink a mixture of warm water with two teaspoons of honey and a few drops of fruit juice. Filariasis is treated with leaves.[3,18,19]

FIGURE 3.3 Fruits and leaves of *Citrus aurantifolia*.

Source: Image by Forest & Kim Starr. https://creativecommons.org/licenses/by/3.0/

3.3.4 *CITRUS LIMON*

Uses in ethnomedicine: Vitamin C in fruit juice stimulates the immune system, preventing colds and flu. It is used to treat scurvy, circulatory, and arteriosclerosis problems, as well as prevent stomach infections. The herb is a natural antibacterial, antirheumatoid, antiseptic, and anti-inflammatory. In addition to being used as an appetizer, the juice is also used in the treatment of vomiting and nausea.[26,44]

FIGURE 3.4 Fruit of *Citrus limon.*

Source: Image by Elena Chochkova. https://creativecommons.org/licenses/by-sa/4.0/

3.3.5 GLYCOSMIS PENTAPHYLLA

Uses in ethnomedicine: Fever and liver complaints are treated with juice from the leaves. Eczema and other skin conditions can be treated externally with a paste made of leaves and ginger. Root decoctions are used to treat facial inflammation.[53,57]

FIGURE 3.5 Fruits and leaves of *Glycosmis pentaphylla.*

3.3.6 *LIMONIA ACIDISSIMA*

Uses in ethnomedicine: Fruits are used to treat diarrhea and dysentery when they are unripe. Patients with heart conditions can benefit from seeds. The leaves are carminative and purgative, which are helpful in dysentery, indigestion, vomiting, and hiccups. As a remedy for venomous insect bites and reptile stings, the fruit pulp is applied externally on the body and prescribed for biliousness. Asthma, tumors, ophthalmia, leucorrhea, diarrhea, and coughs can also be treated with the pulp; it is also useful to treat dysentery and heart disease.[7,45]

FIGURE 3.6 Fruits of *Limonia acidissima.*

Source: Image by Adityamadhav83. https://creativecommons.org/licenses/by-sa/4.0/deed.en

3.3.7 *MURRAYA KOENIGII*

Uses in ethnomedicine: Diarrhea and dysentery are treated with the leaves. Bruises, eruptions, and fever can be treated with crushed leaves, and leaves brewed in decoction are used as remedies for snake bites and bruises. Stomach upset is treated with an infusion of roasted leaves. A mild purgative is derived from the root of the plant, and its juice is used to treat kidney pain. Roots and bark are antimicrobial and cytotoxic; they inhibit tumor growth.[6,21]

FIGURE 3.7 Leaves of *Murraya koenigii.*

Source: Image by Krzysztof Ziarnek, Kenraiz. https://creativecommons.org/licenses/by-sa/4.0/deed.en

3.3.8 MURRAYA PANICULATA

The leaves are traditionally used for dysentery, diarrhea, and toothaches. Drops and powdered leaves are used on fresh cuts after the leaves are boiled and then applied to the cuts. Bodyaches are relieved by eating and applying the bark of ground roots. Cosmetics are made from bark and flowers.[17,35,50]

FIGURE 3.8 Flowers and leaves of *Murraya paniculata.*

Source: Image by Chiring chandan. https://creativecommons.org/licenses/by-sa/4.0/deed.en

3.3.9 *PHELLODENDRON AMURENSE*

The crude extract of plant is used as a natural pesticide. The fruits and leaves are used for skin rashes, GIT disorders, night sweating, and diarrhea.[46,62]

FIGURE 3.9 Fruits and leaves of *Phellodendron amurense.*

Source: Image by Baummapper. https://creativecommons.org/licenses/by-sa/3.0/de/deed.en

3.3.10 *DICTAMNUS DASYCARPUS*

The plant is used externally for liver diseases, skin diseases, reduction of fever heat, and propagation of cancerous cells.[24,36]

FIGURE 3.10 Flowers of *Dictamnus dasycarpus.*

Source: Image by yakovlev.alexey from Moscow, Russia. https://creativecommons.org/licenses/by-sa/2.0/deed.en

3.4 SUMMARY

The Rutaceae family is a diverse group of plants, which provides medicinal materials for traditional medicines. The knowledge of ethnomedicinal values has been transferred from generation to generation, mainly verbally. It is difficult to collect all the ethnomedicinal values of the plants. A modern ethnomedical approach allows for the development of innovative drugs and treatments, as well as preserves medicinal plants by recording how plants have been used for centuries. The members of family Rutaceae have been used in traditional medicines for prevention and treatment of various diseases like diarrhea, abdominal pain, stomachache, etc. This chapter summarizes the research documents on the ethnomedicinal importance of the Rutaceae family. More detailed research is needed to assess the real therapeutic potential of the Rutaceae family.

KEYWORDS

- **citrus**
- ***Clausena***
- **ethnomedicinal importance**
- **plants**
- **Rutaceae**
- **therapeutic**
- **traditional**
- ***Zanthoxylum***

REFERENCES

1. Abirami, A.; Nagarani, G.; Siddhuraju, P. Antimicrobial Activity of Crude Extract of Citrus hystrix and Citrus maxima. *Int. J. Pharma Sci. Res.* **2013,** *4*, 1–5.
2. Afolayan, A. J.; Grierson, D. S.; Mbeng, W. O. Ethnobotanical Survey of Medicinal Plants used in the Management of Skin Disorders among the Xhosa Communities of the Amathole District, Eastern Cape, South Africa. *J. Ethnopharmacol.* **2014,** *153* (1), 220–232.
3. Aibinu, I.; Adenipekun, T.; Adelowotan, T.; Ogunsanya, T.; Odugbemi, T. Evaluation of the Antimicrobial Properties of Different Parts of Citrus Aurantifolia (Lime Fruit) as Used Locally. *Afr. J. Tradit. Complement. Altern. Med.* **2007,** *4* (2), 185.

4. Arbab, I. A.; Abdul, A. B.; Aspollah, M.; Abdelwahab, S. I.; Ibrahim, M. Y.; Ali, Z. A Review of Traditional Uses, Phytochemical and Pharmacological Aspects of Selected Members of Clausena genus (Rutaceae). *J. Med. Plants Res.* **2012,** *6* (38), 5107–5118.
5. Arrieta, J.; Reyes, B.; Calzada, F.; Cedillo-Rivera, R.; Navarrete, A. Amoebicidal and Giardicidal Compounds from the Leaves of Zanthoxylum Liebmannianun. *Fitoterapia* **2001,** *72* (3), 295–297.
6. Balakrishnan, R.; Vijayraja, D.; Jo, S. H.; Ganesan, P.; Su-Kim, I.; Choi, D. K. Medicinal Profile, Phytochemistry, and Pharmacological Activities of Murraya koenigii and Its Primary Bioactive Compounds. *Antioxidants (Basel)* **2020,** *9* (2), 101.
7. Bandara, B. R.; Gunatilaka, A. L.; Wijeratne, E. K.; Adikaram, N. K. B. Antifungal Constituents of Limonia Acidissima. *Planta Medica.* **1988,** *54* (04), 374–375.
8. Bandi, A. K. R.; Lee, D. U.; Tih, R. G.; Gunasekar, D.; Bodo, B. Phytochemical and Biological Studies of Ochna Species. *Chem. & Biodiversity* **2012,** *9* (2), 251–271.
9. Baydoun, S.; Chalak, L.; Dalleh, H.; Arnold, N. Ethnopharmacological Survey of Medicinal Plants used in Traditional Medicine by the Communities of Mount Hermon, Lebanon. *J. Ethnopharmacol.* **2015,** *173*, 139–156.
10. Birjees, M.; Ahmad, M.; Zafar, M.; Nawaz, S.; Jehanzeb, S.; Ullah, F.; Zaman, W. Traditional Knowledge of Wild Medicinal Plants Used by the Inhabitants of Garam Chashma Valley, District Chitral, Pakistan. *Acta Ecologica. Sinica.* **2021.**
11. Borah, M.; Ahmed, S.; Das, S. A Comparative Study of the Antibacterial Activity of the Ethanolic Extracts of Vitex Negundo L., Fragaria vesca L., Terminalia Arjuna and Citrus Maxima. *Asian J. Pharm. & Biol. Res.* **2012,** *2* (3).
12. Brijesh, S.; Daswani, P.; Tetali, P.; Antia, N.; Birdi, T. Studies on the Antidiarrhoeal Activity of Aegle Marmelos Unripe Fruit: Validating its Traditional Usage. *BMC Complement. Altern. Med.* **2009,** *9* (1), 1–12.
13. Chaudhari, S. Y.; Ruknuddin, G.; Prajapati, P. Ethno Medicinal Values of Citrus Genus: A Review. *Med. J. Dr. DY Patil Univ.* **2016,** *9* (5), 560.
14. Chhabra, S. C.; Mahunnah, R. L. A.; Mshiu, E. N. Plants Used in Traditional Medicine in Eastern Tanzania. II. Angiosperms (Capparidaceae to Ebenaceae). *J. Ethnopharmacol* **1989,** *25* (3), 339–359.
15. Choudhary, Y.; Saxena, A.; Kumar, Y.; Kumar, S.; Pratap, V. Phytochemistry, Pharmacological and Traditional Uses of Aegle Marmelos. *Pharm. Biosci. J.* **2017,** 27–33.
16. Dandena, T. *Phytochemical Investigation of Root of Clausena anisata for Antibacterial Activity* (Doctoral dissertation, ASTU), 2018.
17. Dosoky, N. S.; Satyal, P.; Gautam, T. P.; Setzer, W. N. Composition and Biological Activities of Murraya Paniculata (L.) Jack Essential Oil from Nepal. *Medicines* **2016,** *3* (1), 7.
18. Ebana, R. U. B.; Madunagu, B. E.; Ekpe, E. D.; Otung, I. N. Microbiological Exploitation of Cardiac Glycosides and Alkaloids from Garcinia Kola, Borreria Ocymoides, Kola Nitida and Citrus Aurantifolia. *J. Appl. Bacteriol.* **1991,** *71* (5), 398–401.
19. Enejoh, O. S.; Ogunyemi, I. O.; Bala, M. S.; Oruene, I. S.; Suleiman, M. M.; Ambali, S. F. Ethnomedical Importance of Citrus Aurantifolia (Christm) Swingle. *Pharma. Innov.* **2015,** *4* (8, Part A), 1.

20. Gu, R.; Wang, Y.; Long, B.; Kennelly, E.; Wu, S.; Liu, B. ... Long, C. Prospecting for Bioactive Constituents from Traditional Medicinal Plants through Ethnobotanical Approaches. *Biological. Pharm. Bull.* **2014,** *37* (6), 903–915.
21. Handral, H. K.; Pandith, A.; Shruthi, S. D. A Review on Murraya Koenigii: Multipotential Medicinal Plant. *Asian. J. Pharm. Clin. Res.* **2012,** *5* (4), 5–14.
22. Huang, L.; Zhe-Ling, F. E. N. G.; Yi-Tao, W. A. N. G.; Li-Gen, L. I. N. Anticancer Carbazole Alkaloids and Coumarins from Clausena Plants: A Review. *Chin. J. Nat. Med.* **2017,** *15* (12), 881–888.
23. Jadid, N.; Kurniawan, E.; Himayani, C. E. S.; Prasetyowati, I.; Purwani, K. I.; Muslihatin, W.; ... Tjahjaningrum, I. T. D. An Ethnobotanical Study of Medicinal Plants Used by the Tengger Tribe in Ngadisari Village, Indonesia. *Plos One* **2020,** *15* (7), e0235886.
24. Jang, J. S.; Seo, E. G.; Han, C.; Chae, H. B.; Kim, S. J.; Lee, J. D.; Wang, J. H. Four Cases of Toxic Liver Injury Associated with Dictamnus Dasycarpus. *Update* **2010**.
25. Kala, C. P.; Farooquee, N. A.; Dhar, U. Traditional Uses and Conservation of Timur (Zanthoxylum Armatum DC.) through Social Institutions in Uttaranchal Himalaya, India. *Conserv. Soc.* **2005,** *3* (1), 224–230.
26. Klimek-Szczykutowicz, M.; Szopa, A.; Ekiert, H. Citrus Limon (Lemon) Phenomenon—A Review of the Chemistry, Pharmacological Properties, Applications in the Modern Pharmaceutical, Food, and Cosmetics Industries, and Biotechnological Studies. *Plants* **2020,** *9* (1), 119.
27. Kumar, K. S.; Umadevi, M.; Bhowmik, D.; Singh, D. M.; Dutta, A. S. Recent Trends in Medicinal Uses and Health Benefits of Indian Traditional Herbs Aegle Marmelos. *Pharma. Innov.* **2012,** *1* (4).
28. Kunwar, R. M.; Bussmann, R. W. Ethnobotany in the Nepal Himalaya. *J. Ethnobiol. Ethnomedicine* **2008,** *4* (1), 1–8.
29. Lawal, I. O.; Uzokwe, V. N. E.; Rafiu, B.; Afolayan, A. J. Foliar Anatomy of Clausena Anisata (willd.) Hook: A South African Medicinal Plant. *Niger. J. Nat. Prod. Med.* **2018,** *22*, 149–154.
30. Lechaba, N. M.; Schutte, P. J.; Hay, L. The Cardiovascular Effects of Clausena Anisata Extracts in Experimental Hypertension. Doctoral Dissertation, University of Limpopo (Medunsa Campus), 2011.
31. Lennox, S.; Bamford, M.; Wadley, L. Middle Stone Age Wood use 58 000 Years Ago in KwaZulu-Natal: Charcoal Analysis from Two Sibudu Occupation Layers. *South Afr Humanit.* **2017,** *30* (1), 247–286.
32. Liengme, C. A. Plants Used by the Tsonga People of Gazankulu. *Bothalia* **1981,** *13* (3/4), 501–518.
33. Lim, T. K. Clausena Lansium. In *Edible Medicinal and Non-medicinal Plants*; Springer, Dordrecht, 2012; pp 871–883.
34. Lu, Q.; Ma, R.; Yang, Y.; Mo, Z.; Pu, X.; Li, C. Zanthoxylum Nitidum (Roxb.) DC: Traditional Uses, Phytochemistry, Pharmacological Activities and Toxicology. *J. Ethnopharmacol.* **2020,** *260*, 112946.
35. Lv, H. N.; Guo, X. Y.; Tu, P. F.; Jiang, Y. Comparative Analysis of the Essential Oil Composition of Murraya Paniculata and M. Exotica. *Nat. Prod. Commun.* **2013,** *8* (10),

36. Lv, M.; Xu, P.; Tian, Y.; Liang, J.; Gao, Y.; Xu, F.; ... Sun, J. Medicinal Uses, Phytochemistry and Pharmacology of the Genus Dictamnus (Rutaceae). *J. Ethnopharmacol.* **2015,** *171*, 247–263.
37. Mabberley, D. J. Citrus (Rutaceae): A Review of Recent Advances in Etymology, Systematics and Medical Applications. *Blumea-Biodivers., Evolut. Biogeogr. Plants* **2004,** *49* (2–3), 481–498.
38. Maneenoon, K.; Khuniad, C.; Teanuan, Y.; Saedan, N.; Prom-In, S.; Rukleng, N.; ... Wongwiwat, W. Ethnomedicinal Plants Used by Traditional Healers in Phatthalung Province, Peninsular Thailand. *J. Ethnobiol. Ethnomed.* **2015,** *11* (1), 1–20.
39. Milner, P. H.; Coates, N. J.; Gilpin, M. L.; Spear, S. R.; Eggleston, D. S. SB-204900, A Novel Oxirane Carboxamide from Clausena Lansium. *J. Nat. Prod.* **1996,** *59* (4), 400–402.
40. Misra, L. N.; Wouatsa, N. V.; Kumar, S.; Kumar, R. V.; Tchoumbougnang, F. Antibacterial, Cytotoxic Activities and Chemical Composition of Fruits of Two Cameroonian Zanthoxylum Species. *J. Ethnopharmacol.* **2013,** *148* (1), 74–80.
41. Mohanapriya, M.; Ramaswamy, L.; Rajendran, R. Health and Medicinal Properties of Lemon (Citrus Limonum). *Int. J. Ayurvedic Herbal Med.* **2013,** *3* (1), 1095–100.
42. Moshi, M. J.; Kagashe, G. A.; Mbwambo, Z. H. Plants Used to Treat Epilepsy by Tanzanian Traditional Healers. *J. Ethnopharmacol.* **2005,** *97* (2), 327–336.
43. Muthee, J. K.; Gakuya, D. W.; Mbaria, J. M.; Kareru, P. G.; Mulei, C. M.; Njonge, F. K. Ethnobotanical Study of Anthelmintic and Other Medicinal Plants Traditionally Used in Loitoktok District of Kenya. *J. Ethnopharmacol.* **2011,** *135* (1), 15–21.
44. Otang, W. M.; Afolayan, A. J. Antimicrobial and Antioxidant Efficacy of Citrus Limon L. Peel Extracts Used for Skin Diseases by Xhosa Tribe of Amathole District, Eastern Cape, South Africa. *South Afr. J. Bot.* **2016,** *102*, 46–49.
45. Pandavadra, M.; Chanda, S. Development of Quality Control Parameters for the Standardization of Limonia Acidissima L. Leaf and Stem. *Asian Pac. J. Trop. Med.* **2014,** *7*, S244–S248.
46. Park, E. K.; Rhee, H. I.; Jung, H. S.; Ju, S. M.; Lee, Y. A.; Lee, S. H.; ... Kim, K. S. Antiinflammatory Effects of a Combined Herbal Preparation (RAH13) of Phellodendron Amurense and Coptis Chinensis in Animal Models of Inflammation. *Phytother. Res.* **2007,** *21* (8), 746–750.
47. Peng, W.; Fu, X.; Li, Y.; Xiong, Z.; Shi, X.; Zhang, F.; ... Li, B. Phytochemical Study of Stem and Leaf of Clausena lansium. *Molecules* **2019,** *24* (17), 3124.
48. Rahman, S.; Parvin, R. Therapeutic Potential of Aegle Marmelos (L.)-An Overview. *Asian Pac. J. Trop. Dis.* **2014,** *4* (1), 71–77.
49. Ripunjoy, S. Indigenous Knowledge on the Utilization of Medicinal Plants by the Sonowal Kachari Tribe of Dibrugarh District in Assam, North-East India. *Int. Res. J. Biol. Sci.* **2013,** *2* (4), 44–50.
50. Sayar, K.; Paydar, M.; Pingguan-Murphy, B. Pharmacological Properties and Chemical Constituents of Murraya Paniculata (L.) Jack. *Med. Aromat. Plants* **2014,** *3* (4), 1–6.
51. Shah, G. M.; Khan, M. A. Common Medicinal Folk Recipes of Siran Valley, Mansehra, Pakistan. *Ethnobotanical Leaflets.* **2006,** *2006* (1), 5.

52. Sheik, H. S.; Vedhaiyan, N.; Singaravel, S. Evaluation of Central Nervous System Activities of Citrus Maxima Leaf Extract on Rodents. *J. App. Pharm Sci.* **2014,** *4* (9), 77.
53. Shoja, M. H.; Reddy, N. D.; Nayak, P. G.; Srinivasan, K. K.; Rao, C. M. Glycosmis Pentaphylla (Retz.) DC Arrests Cell Cycle and Induces Apoptosis via Caspase-3/7 Activation in Breast Cancer Cells. *J. Ethnopharmacol.* **2015,** *168*, 50–60.
54. Sidahmed, H. M. A.; Vadivelu, J.; Loke, M. F.; Arbab, I. A.; Abdul, B.; Sukari, M. A.; Abdelwahab, S. I. Anti-ulcerogenic Activity of Dentatin from Clausena Excavata Burm. f. Against Ethanol-induced Gastric Ulcer in Rats: Possible Role of Mucus and Anti-oxidant Effect. *Phytomedicine* **2019,** *55*, 31–39.
55. Singh, H.; Dhole, P. A.; Saravanan, R. Unreported Ethnomedicinal Uses of Some Plants in Nuapada District, Odisha, India.
56. Singh, T. P.; Singh, O. M. Phytochemical and Pharmacological Profile of Zanthoxylum Armatum DC.-An Overview, 2011.
57. Sreejith, P. S.; Praseeja, R. J. A Review on the Pharmacology and Phytochemistry of Traditional Medicinal Plant, Glycosmis Pentaphylla (Retz.) Correa. *J. Pharm. Res.* **2012,** *5* (5), 2723–2728.
58. Sripisut, T.; Laphookhieo, S. Carbazole Alkaloids from the Stems of Clausena Excavata. *J. Asian Nat. Prod. Res.* **2010,** *12* (7), 614–617.
59. Srivastava, A. K. Significance of Medicinal Plants in Human Life. In *Synthesis of Medicinal Agents from Plants*. Elsevier, 2018; pp 1–24.
60. Tatsadjieu, L. N.; Ngang, J. E.; Ngassoum, M. B.; Etoa, F. X. Antibacterial and Antifungal Activity of Xylopia Aethiopica, Monodora Myristica, Zanthoxylum Xanthoxyloıdes and Zanthoxylum Leprieurii from Cameroon. *Fitoterapia* **2003,** *74* (5), 469–472.
61. Van Sam, H.; Baas, P.; Keßler, P. J. Traditional Medicinal Plants in Ben En National Park, Vietnam. *Blumea-Biodiversity, Evol. Biogeogr. Plants* **2008,** *53* (3), 569–601.
62. Wang, W.; Li, Q.; Liu, Y.; Chen, B. Ionic Liquid-aqueous Solution Ultrasonic-assisted Extraction of Three Kinds of Alkaloids from Phellodendron Amurense Rupr and Optimize Conditions use Response Surface. *Ultrason. Sonochem.* **2015,** *24*, 13–18.
63. Wang, Y. S.; He, H. P.; Yang, J. H.; Di, Y. T.; Hao, X. J. New Monoterpenoid Coumarins from Clausena Anisum-olens. *Molecules* **2008,** *13* (4), 931–937.
64. Xiu-fang, S. U. Ultrasonic-assisted Extraction Technology of Total Flavonoids from the Nutlets of Clausena Anisum-olefins. *Food Sci. Tech.* **2011,** 02.
65. POWO (2023). "Plants of the World Online. Facilitated by the Royal Botanic Gardens, Kew. Published on the Internet; http://www.plantsoftheworldonline.org/ Retrieved 06 September 2023.

CHAPTER 4

Phytochemistry and Medicinal Uses of the Family Rutaceae

NASIB ZAMAN[1], MUHAMMAD RIZWAN[1], ARSHAD IQBAL[1], ABDUR RAUF[2], YAHYA S. AL-AWTHAN[3,4], and OMAR BAHATTAB[4]

[1]*Centre for Biotechnology and Microbiology, University of Swat, Khyber Pakhtunkhwa, Pakistan*

[2]*Department of Chemistry, University of Swabi, Anbar, Swabi, Khyber Pakhtunkhwa, Pakistan*

[3]*Department of Biology, Faculty of Science, University of Tabuk, Tabuk, Saudi Arabia*

[4]*Department of Biology, Faculty of Science, Ibb University, Ibb, Yemen*

ABSTRACT

The family Rutaceae consists of more than 170 species around the world. Many therapeutic molecules can be found in the Rutaceae family. Citrus fruits contain a variety of phytochemicals with antioxidant properties that are beneficial to human nutrition. Citrus fruits' antioxidant activity is predominantly due to their high vitamin C, polyphenols, and carotenoids content. Many research organizations have conducted extensive research on the possible uses of these bioactive compounds in the cure of diseases such as depression, Alzheimer's disease, and cancer and the treatment of various microbial infections. Their antibacterial, antileishmanial, antifungal, and antiplasmodial potential contribute due to extracted natural

Phytochemical and Pharmacological Investigation of the Family Rutaceae.
Abdur Rauf, Hafiz Ansar Rasul Suleria, & Syed Muhammad Mukarram Shah (Eds.)

products. These species not only have the prospective for activities against pathogenic and resistant microorganisms, but they also show great potential to treat different cancers and metabolic disorders. Research activities on producing bioactive compounds from fruits, leaves, stems, and oils are summarized here. This chapter also contains the information of a variety of different natural products such as alkaloids, flavonoids, and tannins. More detailed studies are required to assess the potential activities of these species. There is a need to optimize production to improve the quality and quantity of biological phytochemicals against various ailments. Pharma technology and new techniques are required to breed improved cultivars with high bioactive compounds, develop a quick propagation strategy to generate this high-demanding species, and grow these for production and other pharmaceutical purposes. New tools for extracting substances from the plant material and ways for verifying their structural elucidation motivate researchers to look into their physiologically active ingredients deeper.

4.1 INTRODUCTION

Rutaceae is a family of evergreen perennial trees and shrubs, with a few herbs and climbers thrown in for good measure. Glandular punctate leaves are simple or complex, exstipulate, alternating or opposite, and sometimes reduced to spines. It has solitary inflorescences, axillary or terminal cymes, panicles, and occasionally simple racemes. Flowers can be bisexual or staminate. Sepals (3-) 4-5, free as well as basally united. Petals (3-) 4-5, imbricate. Stamens are equal to or double the number of petals or more. Filaments are free, or mono- or polyadelphous; stamens are equivalent to or twice the number of petals or more superior; 4–5-locular ovary; styles as many as the carpels, free or joined. A capsule, berry, drupe, or schizocarp is a type of fruit with oblong or reniform seeds. It is a diverse family with 150 genera and about 1200 species (wild and cultivated) found in southeast Asia's tropical and subtropical climates, the Mediterranean, North America, Australia, and southern Africa. The Rutaceae family possesses a strong scent and comprises essential oils. Several genera display xerophytic and hardy features.[1] Citrus fruits seem to be the most popular and important fruit crop in the world. These plants are grown in many regions including tropical and subtropical. Citrus fruits are prized for their flavor, scent, and appealing appearance. Citrus fruits are high in phytochemicals, which are crucial for human nutrition since they contain antioxidants.[2]

Citrus belongs to the Rutaceae family and is one of the most common fruit tree cultivars worldwide. Though *Citrus sinensis* is the most common fruit in this category, it only accounts for roughly 70% of citrus production. Small citrus fruits such as tangerine trees (*Citrus reticulata*), grapefruit trees (*Citrus vitis*), lime trees (*Citrus aurantifolia*), and lemon trees (*Citrus limonum*) are also included in this group.[3] Citrus fruits are known for their aroma, partly because the peel contains flavonoids and limonoids (terpenes), and most of them are rich in juice. Juices contain a lot of citric acids, giving them a unique rich flavor.

Flavonoids and vitamin C are also abundant in them, which include various flavanones and flavonoids.[4] Citrus plants (lime or bitter orange) possess some biologically active substances that can treat colds, sore throat, fever, bronchitis, and asthma. These compounds are also helpful in the treatment of obesity, rheumatoid arthritis, and astringent effects. They can clear oily skin and acne, and are used to cure cuts and insect bites.[5] Leaves of *Citrus maxima* (grapefruit) traditionally produce effects in cholera, epilepsy, and cough. A hot leaf decoction is suitable for reducing swelling and ulcers and for cancer treatment.[6,7] Phytonutrients are mainly natural, biologically active compounds derived from plants that have universal benefits for human health.[8] Citrus provides vitamin C, potassium, folic acid, and pectin, which are active phytochemicals in the body. These phytochemicals decrease the risk of cardiovascular disorder and cancer.[9] By itself, the flavonoid class of compounds has the capacity to adjust the body's response to allergies, carcinogens, and infections. It has been discovered that the flavonoid class of compounds has anti-inflammatory, antibacterial, antifungal, antiallergic, and anticancer effects.[10] Tangerine is a phytochemical substance (polymethoxy) flavonoid present in oranges and citrus peels and can be an anticancer agent.[11] Citrus fruits (lime, lemon, grapefruit, and orange) contain the chemicals limonin and nomilin. The detoxifying glutathione S-transferase enzyme has been demonstrated to be activated by both limonoids. Chemically induced carcinogenesis in lab animals is inhibited by these substances. *Zanthoxylum quinduense* Tul. or chelerythrine, syringaresinol, decarine, 6-acetonyldihydrochelerythrine, evofilin C, p-hydroxybenzaldehyde, vanillic acid; a mixture of sitosterol, stigmasterol, and campesterol; as well as their esters and derivatives were identified in phytochemical analysis. The two chemicals that demonstrated antifungal action were evofilin C and nochelerytrine.[12]

4.2 PHYTOCHEMISTRY OF FAMILY RUTACEAE

Phytochemicals are the chemical compounds or substances found in plants. Phytochemicals are one of the most plentiful and widely dispersed classes of compounds in the plant kingdom. Secondary metabolites are chemicals produced by plants that are not directly engaged in the growing process but act as insect and harmful microbe deterrents. This group includes alkaloids, cyanogenic glycosides, flavonoids, terpenoids, and phenolic compounds.[10,13,14] Plant constituents such as phytochemicals, which play a variety of ecological and physiological roles, are widely distributed. Citrus plants synthesize low molecular phenolic acids (hydroxybenzoic and hydroxycinnamic acids), stilbenes, terpenoids, acetophenones, flavonoids, and tannins.[14–17] Quinoline alkaloids, flavonoids, coumarins, sesquiterpenes, limonoids, and steroids are among the more than 170 chemical substances discovered in the *Dictamnus* genus. The active ingredients in these species are limonoids and quinoline alkaloids, which have anticancer, anti-inflammation, and antimicrobial properties. Quinoline alkaloids and limonoids could be used to discriminate between different *Dictamnus* species as quality control markers. Certain reports on toxic hepatitis and the phototoxic effect of *Dictamnus* species, on the other hand, advised that more research be done to verify their findings. Quinoline alkaloids and limonoids were identified as the most promising bioactive components showing the potential in neuroprotective, anti-inflammatory, cytotoxic, and antibacterial effects. These two compounds have triggered a lot of attention, and they may have a lot of potential as new medication lead compounds.[18] Some of the important phytochemicals have been shown in Tables 4.1 and 4.2.

4.2.1 PHENOLIC COMPOUNDS PRESENT IN RUTACEAE AND THEIR BIOLOGICAL APPLICATIONS

Citrus fruits are high in flavonoids and phenolic acids, which are phenolic compounds. Since several epidemiological studies have linked the intake of polyphenol-rich foods and beverages to a lower risk of cardiovascular disease, stroke, and some cancers, the phenolic chemicals found in citrus fruits have attracted more attention in recent years. These compounds are supposed to play a significant role in the antioxidant capacity of citrus fruits.[19–21]

Citrus fruits are the major source of nutrients like vitamin C, flavonoids, and folate, which are proposed to be accountable for the treatment of cancer and some degenerative diseases.[21,22] There are more than 6000 molecules in the flavonoid's class, which are divided into subgroups such as flavanones, flavanols, flavones, anthocyanins, and isoflavones. These can be found in abundance in flowers, fruits, and leaves, and they serve a variety of purposes.[21,23] The members of citrus fruits are high in flavanones, a type of flavonoid that is naturally synthesized in the fruit and plays a role in the prevention of a variety of human diseases.[21,24,25] Some of them can be found as glycosides or aglycones. The most important flavanones are naringenin and hesperetin in the aglycone form, and neohesperidosides and retinoids in the glycoside form.[21,26,27] Flavones are isolated primarily from citrus essential oils that contain diosmin, apigenin, luteolin, diosmetin, and tangeretin, the main flavones found in citrus. Flavanols are a type of flavonoid that have anti-inflammatory and antitumor capabilities, as well as free radical scavenging, tumor mitotic cycle modification, gene expression regulation, and antiangiogenesis effects.[21,28,29]

4.2.2 ALKALOIDS PRESENT IN RUTACEAE AND THEIR BIOLOGICAL IMPORTANCE

Aegle marmelos, the therapeutic usefulness of *Correa* as a constituent of various traditional systems for the cure of a wide range of human illnesses, has been identified. Many biologically active secondary metabolites were detected in *A. marmelos* crude extract, with the highest concentrations of flavonoids, alkaloids, and phenols. The plant extracts have been tested against pathogenic bacteria, with aqueous extract having the strongest inhibitory activity against *S. epidermidis* and methanolic extract having the maximum inhibitory activity against *S. aureus*. Ethanolic extracts were observed to have antibacterial properties against *K. pneumoniae*, *E. aerogenes*, and *S. epidermidis.*[30] A list of different alkaloids is shown in Table 4.1.

4.2.3 FLUOROQUINOLONE ALKALOIDS PRESENT IN RUTACEAE AND THEIR BIOLOGICAL ACTIVITIES

These compounds are extracted from root and bark of *Dictamnus dasycarpus* Turcz (Rutaceae). The plant material has been used in folk medicine for a long time for their biological activities. *D. dasycarpus* Turcz. is used in traditional Chinese medicine; this plant's derivatives are also used to treat rheumatoid arthritis, jaundice, skin disorders, and cough [DCTM, 1996]. In vitro activities against *Leishmania* and *P. falciparum* were also investigated by Basco and colleagues in 1994,[31] and similar results were also reported by Fournet and his colleagues in 1994.[32] In addition, dictamnine was also isolated from *D. dasycarpus* Turcz and proved antifungal properties.[33] Many pharmacological studies were conducted on dichloromethane extracts containing decamine to find the toxic effects of alkaloid. Anticancer capabilities of their derivatives have been proven. They have the ability to halt the growth of lung cancer, breast cancer cells, and cancers of the central nervous system.[34] In addition, in a study conducted[35] by Emam in 2010, he made a new fluoroquinolone alkaloid discovery. *Ruta chalepensis* L.'s 5-(1,1-dimethylallyl)-8-hydroxyfuran has also been demonstrated to have antifungal activities. It was found that chloroform extracts from this species' leaves were effective against three fungi (*Sclerotium rolfsii*, *Fusarium solani*, and *Rhizoctonia solani*). Potatoes, tomatoes, and beets were all affected by these fungi. The impact of a methanol extract of *Ruta chalepensis* L. aerial parts on *Culex pipiens* larvae was researched by other researchers.[34] The extracts of *Ruta chalepensis* L. are efficient against larvae but pose little or no damage to animals or humans. Scopoletin, kokusaginine, pseudane IX, γ-fagarine, chalepin, and arborinine were isolated from *Ruta angustifolia* Pers. leaves by Wahyunia and his study group.[36] In cell cultures, these chemicals displayed antiviral activity against HCV.

Isolated compounds including isofagadine, macurin, kokusaginin, tecleaverdoornin, and buegenine extracted from the aerial portions of *Zanthoxylum buesgenii* (Engl.) showed activity against breast cancer, colon cancer, leukemia, drug-resistant subline, glioblastoma, and liver cancer on the tested cancer cell lines.[36] Isofagadine and kokusaginine have shown to be more effective than doxorubicin.[36]

In 2013, Tavares[37] tested the alkaloid extracts from *Zanthoxylum rhoifolium* Lam. for antimicrobial activity, and his experiment proved broad-spectrum antibacterial activity. Three alkaloids, including γ-fagarine and isohaplopine, were also isolated and used against pathogenic microorganisms. All the compounds were screened against gram (+ve) bacteria including *Bacillus cereus*, *S. aureus*, *Bacillus subtilis*, *Streptococcus pyogenes*, *S. epidermidis*, *Enterococcus*, and *Enterobacter aerogenes*. The compounds were also effective against gram (−ve) bacteria including *K. pneumoniae*, *E.coli*, *P. aeruginosa*, *Enterobacter cloacae*, *Salmonella typhimurium*, *Shigella sonnei*, *Burkholderia cepacia*, *C. albicans*, *Morganella*, *Candida tropicalis*, and *Candida krusei*.[37] The leaves of *Evodia lepta* have also yielded three leptanoines including dictamnine, melineurine, and 7-hydroxydictamnine.[38]

Alkaloids obtained from citrus plants have an importance in treatment of different kinds of diseases, such as depression, related to serotonin neurotransmission.[39] Alkaloid compounds, skimmianine, maculine, kokusaginine, and flindersiamine, were also obtained from the stems and leaves of *Raulinoa echinata*, and showed similar results in treating depression. Furthermore, quinolone derivatives (1-methyl-2-n-nonyl-4-quinolone, 1-methyl-2-phenyl-4-quinolone, and 2-n-nonyl-4-quinolone) were also extracted from the same plant.[40] These alkaloids have shown an inhibitory biological role against *Leucoagaricus gongylophorus* (fungus). Through in vitro studies, quinolinone alkaloids have revealed activity against *Leishmania*, *Trypanosoma cruzi*, and *Plasmodium falciparum*.[40]

In a previous study conducted by Cabral and research colleagues,[41] extracted alkaloids from the stems of *Conchocarpus fontanesianus* displayed an inhibitory activity on acetylcholinesterase. More activity was shown by skimmianine compound. Few phytochemicals from Brazilian plant were also used in the treatment of Alzheimer's disease. The results of other research works[42] showed that the eptomerine and skimmianine (alkaloids) are effective in anticholinesterase activity.

Mwangi et al.[43] tested crude methanol extract, and they found chemicals identified from *Teclea trichocarpa* leaves. These bioactive compounds were effective against *Trypanosoma cruzi*, *P. falciparum*, and *Leishmania donovani*. With the lowest inhibitory concentration, skimmianine showed potent activity against *T. brucei* and *T. cruzi* and cytotoxic activity against L-6 cells. This study showed that furanoquinoline alkaloids are potential antiprotozoal compounds.[43]

TABLE 4.1 Biological Activities of Many Alkaloids Identified from the Rutaceae Family.

Phytochemical	Source	Part used	Activity	References
Dictamnine	*D. dasycarpus* Turcz	Roots and stems	Antifungal and anticancer	33,34
5-(1,1-dimethylallyl)-8-hydroxyfuro[2-3-b] quinoline	*Ruta chalepensis* L.	Aerial parts	Activity against *Culex pipiens* larvae	34,35
Furochinoline alkaloids	*Evodia lepta*	Leaves	Inhibitory activity against acetylcholinesterase (AChE)	38
Scopoletin, γ-fagarine, chalepin, and kokusaginine	*Ruta angustifolia* Pers	Leaves	Anti-HCV	44
Isofagadine, macurin, and kokusaginine	*Zanthoxylum buesgenii* (Engl.)	Aerial parts	Anticancer	35,36
1-methyl-2-n-nonyl-4-quinolone, 2-n-nonyl-4-quinolone, and 1-methyl-2-phenyl-4-quinolone (quinolone derivatives)	*Raulinoa echinata*	Stems and leaves	Activities against *Leucoagaricus gongylophorus* (fungus), malarial, and Leishmanial parasites	40
Skimmianine, γ-fagarine, and isohaplopine	*Zanthoxylum rhoifolium* Lam.	Bark tissues	Broad-spectrum antimicrobial activities	37,45
Melineurine and 7-hydroxydictamnine	*Evodia lepta*	Leaves	Inhibitory activity against acetylcholinesterase (AChE)	38
Maculine, kokusaginine, and flindersiamine	*Raulinoa echinata*	Stems and leaves	Antidepression activity	40,45
Alkaloids	*Conchocarpus fontanesianus*	Stems	Inhibitory activity against acetylcholinesterase	41
Leiokinine A, leptomerine, kokusaginine, skimmianine, maculine, and flindersiamine	*Esenbeckia leiocarpa* Engl.		Inhibitory activity toward AChE	42

TABLE 4.1 *(Continued)*

Phytochemical	Source	Part used	Activity	References
Alkaloids	Brazilian vegetable sp.	Stems	Activity against Alzheimer's disease	41
Acridone, alkaloids, a furoquinoline alkaloid, and triterpenoids	*Teclea trichocarpa* (Engl.)	Leaves	Activity and cytotoxicity against parasitic protozoa	43
Hydroxy-alpha-sanshool (HAS)	*Z. bungeanum*	Roots and leaves	Produce tingling and numbing sensation	46

4.2.4 ALKYLAMIDES

The most distinctive chemicals are alkylamides, which cause a strong numbing sensation in the mouth. These compounds have been isolated from *Zanthoxylum bungeanum* plant. More than 25 alkylamides have been extracted and identified so far. Due to the presence of two or more conjugated double bonds, they are frequently highly unsaturated and have a distinctive taste. The active element chiefly accountable for the characteristic tingling sensation generated by the pericarps of *Z. bungeanum* is hydroxy-alpha-sanshool (HAS), which has four double bonds in the cis-configuration. In 1982, Yasuda and his colleagues isolated HAS from the pericarps of *Z. bungeanum* and recognized it.[46]

4.2.5 OTHER ALKALOIDS PRESENT IN RUTACEAE FAMILY

Other alkaloids have been identified from *Z. bungeanum* in addition to the alkylamides. There have been eight alkaloids isolated from the roots of *Z. bungeanum*. Des-N-methylchelerythrine, zanthobungeanine, L-N-acetylanonanine, 11-methoxychelerythrine, arnothianamide, and skimmianine were identified in 1981[47,48]; in 1984, the pericarps of *Z. bungeanum* were also discovered to contain haplopine and kokusaginine.[47,48]

4.2.6 ARBORINE AND SKIMMIANINE

Arborine and skimmianine were identified as the isolated compounds from *Glycosmis pentaphylla* (Rutaceae), with the lowest MIC and MBC values against bacteria. They demonstrated considerable bactericidal activity against multidrug-resistant *Staphylococcus aureus*, and in vitro kinetic and protein leakage assays confirmed the antimicrobial action. Arborine and skimmianine treatment caused significant morphological alterations in MDR *S. aureus*, including uneven cell surfaces and shapes, cell shrinkage, and cell membrane damage.[49]

Skimmianine is a furoquinoline alkaloid discovered in the flowering plant *Skimmia japonica* that belongs to the Rutaceae family. It's also an effective inhibitor of acetylcholinesterase (AChE).[50] Arborine is an alkaloid molecule that is a bioactive compound isolated from *Glycosmis arborea* by Chakravarti and his research group[51] and also from *Glycosmis pentaphylla*,[49] which are members of the Rutaceae family. The isolated drug had a considerable antibacterial impact toward MDR *S. aureus*, with the lowest MBC and MIC values; protein leakage assays and in vitro killing kinetic assay also confirmed the antimicrobial activity.[49]

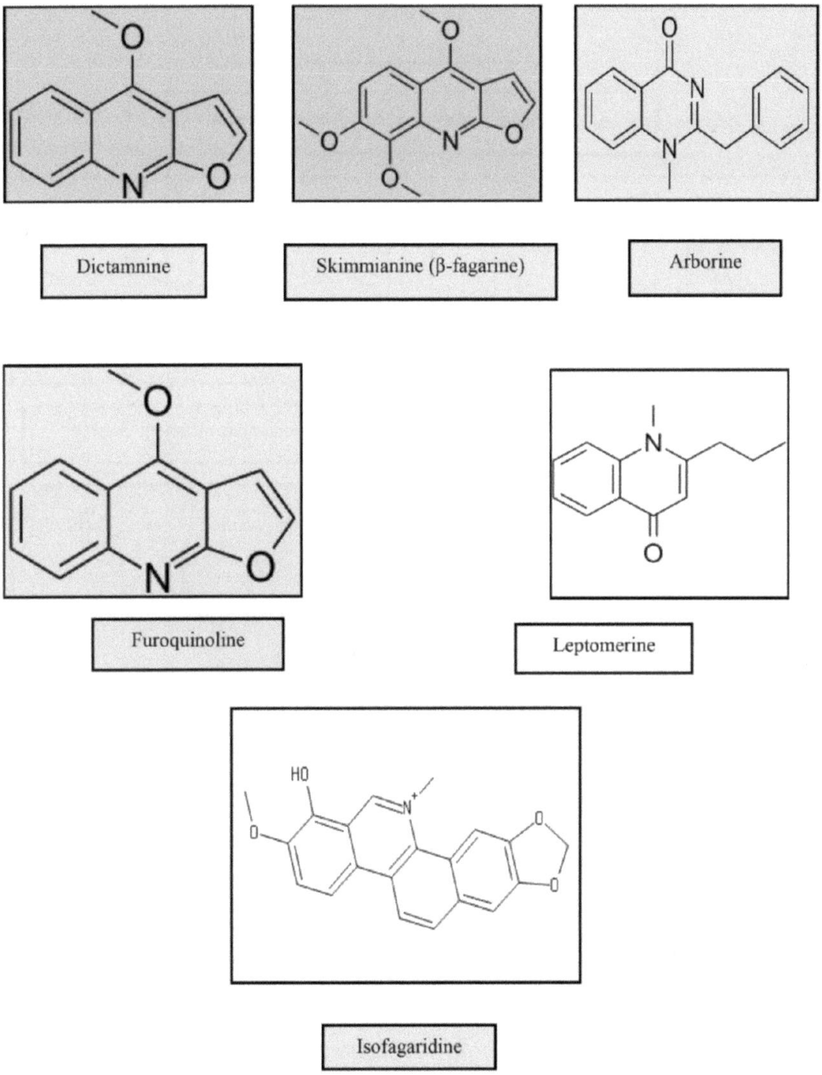

FIGURE 4.1 Different alkaloid compounds isolated from Rutaceae family.

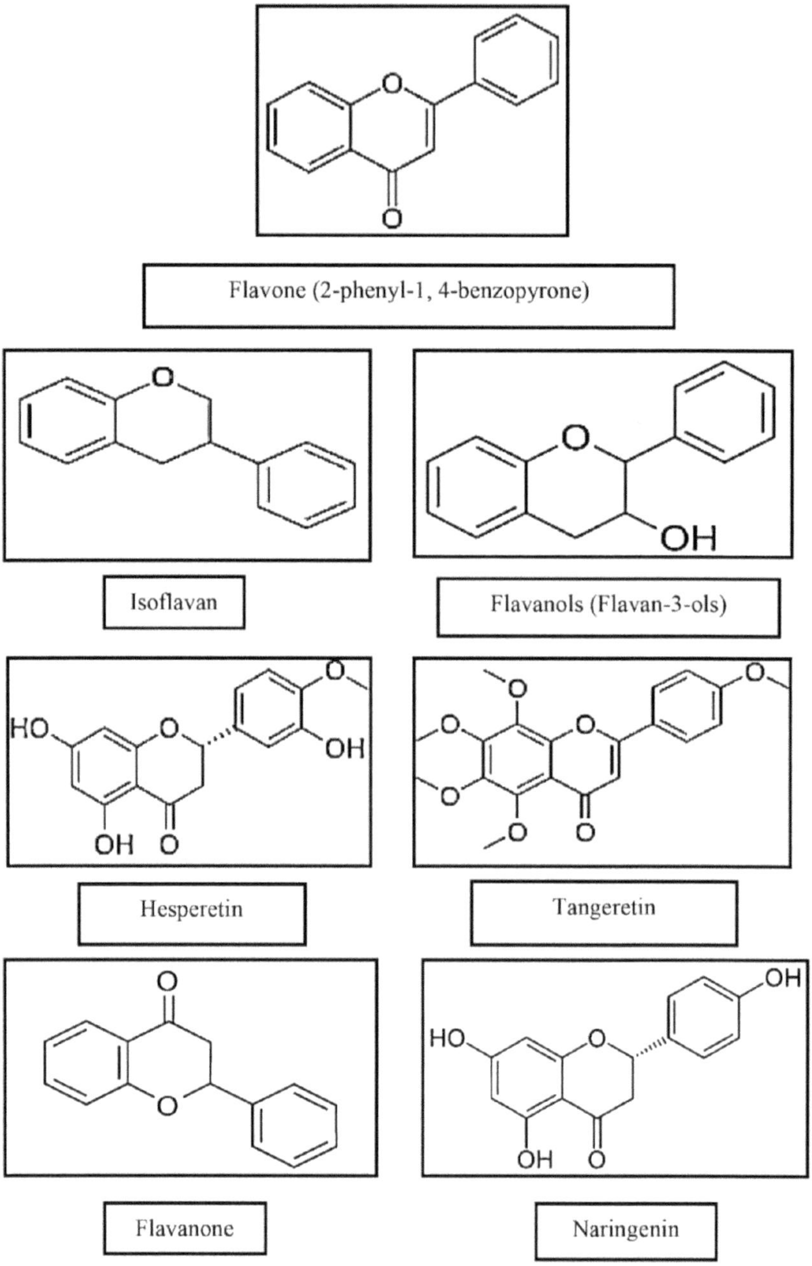

FIGURE 4.2 Flavonoids extracted from different parts of Rutaceae family.

4.2.7 TANNIN

The highest concentrations of tannin have been reported in the leaves of *C. aurantifolia* (1.440.02%) and *C. limon* (1.300.02%). Furthermore, large amounts of tannin were recorded in the roots of *C. aurantifolia* (0.820.04%) and *C. limon* (0.840.04%), as well as the leaves (0.83 ± 0.05%) and the root (0.780.04%) of *C. paradisii*. The astringency of the *Citrus* species could be attributed to their high tannin content. Tannin is thought to have astringent effects.[52,53] As a result of binding to salivary proteins, it produces astringency, which humans recognize as a flavor.[54] This could be because the leaves have high tannin content. *Citrus sinensis* peels have a high degree of tannin (0.900.02%) as well.[52]

4.2.8 FLAVONOIDS

Poncirin, rhoifolin, naringin, and marmesin are four flavonoid compounds identified from *Poncirus trifoliata* that showed potential activities against *Aedes aegypti*. These chemicals have the potential to be exploited as agents in the production of a commercial mosquitocidal medicine that may be used as a substitute for synthetic chemicals in integrated vector control applications.[55] The extract of *Poncirus trifoliata* was used to obtain two phytochemicals, which were identified as β-sitosterol and 2-hydroxy-1,2,3-propanetricarboxylic acid 2-methyl ester (HPCME). Oxygen radical absorbance capacity (ORAC) measurements were used to assess the anti-oxidant capacity of HPCME, β-sitosterol, and Trolox. Furthermore, these two chemicals were tested on a human colon cancer cell line (HT-29) for their ability to inhibit cancer cell proliferation and apoptotic activities.[56] Almost all citrus fruits contain 7-O-glycosylflavanones that are the most plentiful flavonoids.[57] Table 4.2 shows a list of flavonoids extracted from Rutaceae.

4.2.9 HESPERIDIN AND RUTIN

The antioxidant and hypoglycemic properties of *Ruta chalepensis* L. leaves have been documented using DPPH methods. β-carotene bleaching and metal chelating potency studies were used to examine the antioxidant potential. A carbohydrate-hydrolyzing enzyme inhibition procedure was used to determine the hypoglycemic effects. Hesperidin and rutin content were both high in *R. chalepensis*, with values of 591.9 and 266.7 mg/g, respectively. In a concentration-dependent manner, the extract inhibited both amylase and glucosidase enzymes (IC_{50} value of 16.9 g/mL). α-amylase (IC50 = 69.0 g/mL) was the enzyme with the highest activity.[58]

TABLE 4.2 Activities of the Flavonoids and Other Phytoconstituents Identified from Various Extracts of Family Rutaceae.

Compound name	Plant source	Activity	References
Flavonoids (poncirin, rhoifolin, naringin, and marmesin)	*Poncirus trifoliata*	Mosquitocidal against *Aedes aegypti*	55
β-sitosterol and 2-hydroxy-1,2,3-propanetricarboxylic acid 2-methyl ester (HPCME)	*Poncirus trifoliata*	Apoptotic and anticancer activities against the human colon cancer cell line	56
Hesperidin and rutin	*Ruta chalepensis* L.	Antioxidant and hypoglycemic properties	32
Flavonoids and phenols	Aegle marmelos Correa (Rutaceae)	Antibacterial activity against *S. epidermidis*, K. pneumoniae, and S. epidermidis	30
5-dihydroxy-6-methyl, 4H-pyran-4-one, 2,3-dihydro-3,	Aegle marmelos	Anti-inflammatory and antiproliferative	30
Esters (cinnamic acid, 4-hydroxy-3-methoxy-, methyl ester)	Aegle marmelos	Antimicrobial, antioxidant, and antiviral	30
3,7,11,15-tetramethyl-2-hexadecen-1-ol (Phytol)	Aegle marmelos	Antimicrobial, anticancer, anti-inflammatory, and diuretic properties	30
Stigmasta-5,22-dien-3-ol	Aegle marmelos	Antioxidant, anti-inflammatory, antiarthritic, antiasthmatic, and diuretic	30
Vitamin E	Aegle marmelos	Antileukemic, antidermatitic, antitumor, anticancer, and hepatoprotective activity	30
Evofilin C and nochelerytrine	*Zanthoxylum quinduense* Tul	Antifungal activities against *Fusarium oxysporum*	12

TABLE 4.2 *(Continued)*

Compound name	Plant source	Activity	References
Poncirin, rhoifolin, naringin, and marmesin	*Poncirus trifoliata*	Potential activities against *Aedes aegypti*	55
β-sitosterol and 2-hydroxy-1,2,3-propanetricarboxylic acid 2-methyl ester (HPCME)	*Poncirus trifoliata*	Anticolon cancer	56
Tannin	*Citus aurantifolia* and *Citrus limon*	Astringent effects	52,53
Hydroxy-alpha-sanshool (HAS)	*Z. bungeanum*	Use for the tingling and numbing sensation	46
5-(1,1-dimethylallyl)-8-hydroxyfuro [2-3-b] quinolone	*Ruta chalepensis* L	Antifungal activity	35,45
Skimmianine, γ-fagarine, and isohaplopine	*Zanthoxylum rhoifolium*	Antimicrobial activity	45,59

4.3 SUMMARY

Many therapeutic molecules can be found in the Rutaceae family. Many research organizations have conducted extensive research; so, these natural substances could be used to treat diseases including Alzheimer's, cancer, and depression, and in the treatment of various microbial infections. Their antibacterial, antileishmanial, antiplasmodial, and antifungal characteristics contribute due to extracted natural products. Family Rutaceae consists of more than 170 species around the world. These species not only have the potential for activities against pathogenic and resistant microorganisms, but they also show great potential to treat different cancers and metabolic disorders. Citrus fruits contain a variety of phytochemicals with antioxidant properties that are beneficial to human nutrition. Citrus fruits' antioxidant activity is predominantly due to their high vitamin C, polyphenols, and carotenoids content. Research activities on producing bioactive compounds from the fruits, leaves, and stems are summarized. This chapter also represents the resources of knowledge of other multiple natural products such as alkaloids, flavonoids, and tannins. More detailed studies are required to assess the potential activities of these species. There is a need to optimize production to improve the quality and quantity of biological phytochemicals against various ailments. Pharma technology and advanced new techniques are needed to produce more effective bioactive compounds from natural plant sources for the production and other pharmaceutical uses. New technologies for extracting chemicals from medicinal herbs, as well as methods for validating their structures encourage researchers to dig further into their biologically active elements.

ACKNOWLEDGMENTS

I would like to thank Dr. Abdur Rauf, Assistant Professor of Department of Chemistry, University of Swabi, Pakistan. I also present my thanks to all the authors of this book.

KEYWORDS

- **anticancer activities**
- **antimicrobial activities**
- **bioactive compounds**
- **phytochemistry**
- **Rutaceae**

REFERENCES

1. Din, H.-U.; Ghazanfar, S. A. *Flora of Pakistan:* Rutaceae, Department of Botany, University of Karachi; 1980.
2. Sdiri, S.; Salvador, A.; Farhat, I.; Navarro, P.; Besada, C. Influence of Postharvest Handling on Antioxidant Compounds of Citrus Fruits. Citrus: Molecular Phylogeny, Antioxidant Properties and Medicinal Uses. **2014,** 73–94.
3. Okwu, D.; Emenike, I. Nutritive Value and Mineral Content of Different Varieties of Citrus Fruits. *J Food Tech.* **2007,** *5* (2), 105–108.
4. Katrine, B. *The Health Benefits of Citrus Fruits*. Horticultural Australia Ltd.: Sydney, Australia. 2003; pp 2–3.
5. Joy, P.; Mathew, J.; Skaria, B. *Aromatic and Medicinal Plants Research Station*. Kerala Agricultural University: India; 1998.
6. Balamurugan, P.; Rajkumar, A.; Prasad, M. Comparative Phytochemical Analysis of Rutaceae Family (Citrus Species) Extracts. *Int. J. Sci.* **2014,** 148–150.
7. Kirtikar, K.; Basu, B. Eds. *Indian Medicinal Plants*. International Book Distributors: Dehradun, India, Vol. 1, 2005; pp 478–479.

8. Zhao, J. Ed. Nutraceuticals, Nutritional Therapy, Phytonutrients, and Phytotherapy for Improvement of Human Health: A Perspective on Plant Biotechnology Application. *Recent Pat. Biotechnol.* **2007,** *1* (1), 75–97.
9. Craig, W. J. Phytochemicals: Guardians of Our Health. *J. Am. Diet Assoc.* **1997,** *97* (10), 199–204.
10. Okwu, D. Phytochemicals and Vitamin Content of Indigenous Spices of South Eastern. *Nigeria J. Sustain Agri. Environ.* **2004,** *6* (1), 30–37.
11. Orlikowski, L. Effect of Grapefruit Extract on Development of Phytophthora Cryptogea and Control of Foot Rot of Gerbera. *J. Plant Prot. Res.* **2001,** *41* (3).
12. Patiño L.; O. J.; Cuca Suárez, L. E. Chemical Constituents of the Wood from Zanthoxylum Quinduense Tul. (Rutaceae). *Química Nova.* **2010,** *33*, 1019–1021.
13. Ullah, A.; Hassan, S.; Khan, M. I.; Rizwan, M.; Ullah, Z.; Shah, M. Antioxidant, Phytotoxic and Cytotoxic Activity of Methanolic Extract of Trigonella Foenum-graecum. *J. Coast Life Med.* **2016,** *4* (5), 386–389.
14. Rizwan, M.; Khan, A.; Nasir, H.; Javed, A.; Shah, S. Z. Phytochemical and Biological Screening of Berberis Aristata. *Advancements Life Sci.* **2017,** *5* (1), 01–07.
15. Yano, M.; Kawaii, S.; Tomono, Y.; Katase, E.; Ogawa, K. Quantification of Flavonoid Constituent in Citrus Fruits. *J. Agri. Food Chem.* **1999,** *47*, 3565–3571.
16. Del-Rio, A. O. B. G.; Obdululio, B. G.; Casfillo, J.; Marin, F. G, Ortuno, A. Uses and Properties of Citrus Flavonoids. *J. Agri. Food Chem.* **1997,** *45* (31), 4505–4515.
17. Rapisarda, P.; Tomaino, Lo Cascio, A.; Bonina, R.; De Pasquale, F.; Saija, A. Antioxidant Effectiveness as Influenced by Phenolic Content of Fresh Orange Juices. *J. Agri. Food Chem.* **1999,** *47* (11), 4718–4723.
18. Lv, M.; et al. Medicinal Uses, Phytochemistry and Pharmacology of the Genus Dictamnus (Rutaceae). *J. Ethnopharmacol.* **2015,** *171*, 247–263.
19. Wang, H.; Cao, G.; Prior, R. L. Total Antioxidant Capacity of Fruits. *J. Agri. Food Chem.* **1996,** *44* (3), 701–705.
20. Gardner, P. T.; et al. The Relative Contributions of Vitamin C, Carotenoids and Phenolics to the Antioxidant Potential of Fruit Juices. *Food Chem.* **2000,** *68* (4), 471–474.
21. Hayat, K. Citrus Molecular Phylogeny Antioxidant Properties and Medicinal Uses. *Nova Sci.* **2014,** *3*, 235.
22. Ejaz, S.; et al. Limonoids as Cancer Chemopreventive Agents. *J. Agri. Food Chem.* **2006,** *86* (3), 339–345.
23. Harborne, J. B.; Williams, C. A. Advances in Flavonoid Research Since 1992. *Phytochemistry* **2000,** *55* (6), 481–504.
24. Nogata, Y.; et al. Flavonoid Composition of Fruit Tissues of Citrus Species. *Biosci. Biotechnol. Biochemistry* **2006,** *70* (1), 178–192.
25. Yao, L. H.; et al. Flavonoids in Food and Their Health Benefits. *Plant Foods Hum. Nutr* **2004,** *59* (3), 113–122.
26. Gionfriddo, F.; Postorino, E.; Bovalo, F. I Flavanoni Glucosidici Del Succo Di Bergamotto. *Essenze e Derivati Agrumari* **1986,** *66* (4), 404–416.
27. Macheix, J.-J.; Fleuriet, A.; Billot, *J. Fruit Phenolics*; CRC Press; 2018.
28. Hayashi, A.; Gillen, A. C.; Lott, J. R. Effects of Daily Oral Administration of Quercetin Chalcone and Modified Citrus Pectin on Implanted Colon-25 Tumor Growth in Balb-c Mice. *Altern. Med. Rev.* **2000,** *5* (6), 546–552.

29. Robards, K.; Antolovich, M. Analytical Chemistry of Fruit Bioflavonoids A Review. *Analyst* **1997,** *122* (2), 11–34.
30. Mujeeb, F.; Bajpai, P.; Pathak, N. Phytochemical Evaluation, Antimicrobial Activity, and Determination of Bioactive Components from Leaves of Aegle Marmelos. *Bio. Med. Res. Int.* **2014,** 2014.
31. Basco, L. K.; et al. In Vitro Activities of Furoquinoline and Acridone Alkaloids Against Plasmodium Falciparum. *Antimicrob. Agents Chemother.* **1994,** *38* (5), 1169–1171.
32. Fournet, A.; et al. Antiprotozoal Activity of Quinoline Alkaloids Isolated from Galipea Longiflora, a Bolivian Plant Used as a Treatment for Cutaneous Leishmaniasis. *Phytother. Res.* **1994,** *8* (3), 174–178.
33. Wang, L.; et al. Antifungal Effect of Three Natural Products on the Genetic Substance of Saccharomyces Cerevisiae GL7 and Prototheca Wickerhamii. *Yao Xue Xue Bao.* **2000,** *35* (11), 860–863.
34. Abdel-Sattar, E.; et al. Evaluation of Some Medicinal Plants in Controlling Culex Pipiens. *J. Egypt Soc. Parasitol.* **2014,** *44* (3), 771–778.
35. Emam, A.; Eweis, M.; Elbadry, M. A New Furoquinoline Alkaloid with Antifungal Activity from the Leaves of Ruta Chalepensis L. *Drug Discoveries* **2010,** 399.
36. Sandjo, L. P.; et al. Cytotoxic Benzophenanthridine and Furoquinoline Alkaloids from Zanthoxylum Buesgenii (Rutaceae). *Chem. Cent. J.* **2014,** *8* (1), 1–5.
37. Tavares, L.d.C.; et al. Structure-activity Relationship of Benzophenanthridine Alkaloids from Zanthoxylum Rhoifolium Having Antimicrobial Activity. *PloS One* **2014,** *9* (5), 97000.
38. Sichaem, J.; et al. New furoquinoline Alkaloids from the Leaves of Evodia Lepta. *Fish Physiol. Biochem.* **2014,** *92*, 270–273.
39. Cheng, J. T.; Chang, T. K. and. Chent, I. S. Skimmianine and Related Furoquinolines Function as Antagonists of 5-hydroxytryptamine Receptors in Animals. *J. Auton. Pharmacol.* **1994,** *14* (5), 365–374.
40. Biavatti, M. W.; et al. Biological Activity of Quinoline Alkaloids from Raulinoa Echinata and X-ray Structure of Flindersiamine. *J. Braz. Chem. Soc.* **2002,** *13*, 66–70.
41. Cabral, R. S.; et al. Anticholinesterase Activity Evaluation of Alkaloids and Coumarin from Stems of Conchocarpus Fontanesianus. *Revista Brasileira de Farmacognosia* **2012,** *22* (2), 374–380.
42. Cardoso-Lopes, E. M.; et al. Alkaloids from Stems of Esenbeckia Leiocarpa Engl. (Rutaceae) as Potential Treatment for Alzheimer Disease. *Molecules* **2010,** *15* (12), 9205–9213.
43. Mwangi, E.; et al. Antiprotozoal Activity and Cytotoxicity of Metabolites from Leaves of Teclea Trichocarpa. *J. Med. Plants Res.* **2010,** *4* (9), 726–731.
44. Wahyuni, T. S.; et al. Inhibition of Hepatitis C Virus Replication by Chalepin and Pseudane IX Isolated from Ruta Angustifolia Leaves. *Fitoterapia* **2014,** *99*, 276–283.
45. Adamska-Szewczyk, Glowniak, A.; K.; Baj, T. Furochinoline Alkaloids in Plants from Rutaceae Family-A Review. *Curr. Issues Pharm. Med. Sci.* **2016,** *29* (1), 33–38.
46. Yasuda, I.; Takeya, K.; Itokawa, H. Distribution of Unsaturated Aliphatic Acid Amides in Japanese Zanthoxylum Species. *Phytochemistry* **1982,** *21* (6), 1295–1298.
47. Chen, S. Study on the Chemical Composition of Pericarps of Zanthoxylum Bungeanum. *J. Nanjing Univ. Tradit. China Med.* **1984,** *16*, 2–4.

48. Zhang, M.; et al. Zanthoxylum Bungeanum Maxim. (Rutaceae): A Systematic Review of Its Traditional Uses, Botany, Phytochemistry, Pharmacology, Pharmacokinetics, and Toxicology. *Int. J. Mol. Sci.* **2017,** *18* (10), 2172.
49. Murugan, N.; et al. Glycosmis pentaphylla (Rutaceae): A Natural Candidate for the Isolation of Potential Bioactive Arborine and Skimmianine Compounds for Controlling Multidrug-Resistant Staphylococcus aureus. *Frontiers Public Health* **2020,** *8*, 176.
50. Yang, Z.-D.; et al. Skimmianine, A Furoquinoline Alkaloid from Zanthoxylum Nitidum as a Potential Acetylcholinesterase Inhibitor. *Med Chem Res.* **2012,** *21* (6), 722–725.
51. Chakravarti, D.; et al. Alkaloids of Glycosmis Arborea—II: Structure of Arborine. *Tetrahedron* **1951,** *16* (1–4), 224–250.
52. Ezeabara, C. A.; et al. Determination of Tannin Content in Various Parts of Six Citrus Species. *J Scientific Res. Rep.* **2014,** 1384–1392.
53. Okwu, D.; Okwu, M. Chemical Composition of Spondias Mombin Linn Plant Parts. *J. Sustain. Agri. Environ.* **2004,** *6* (2), 140–147.
54. Morton, J. F. *Fruits of Warm Climates*; 1987.
55. Rajkumar, S.; Jebanesan, A. Bioactivity of Flavonoid Compounds from Poncirus Trifoliata L. (Family: Rutaceae) Against the Dengue Vector, Aedes Aegypti L. (Diptera: Culicidae). *Parasitology Res.* **2008,** *104* (1), 19–25.
56. Jayaprakasha, G.; et al. Inhibition of Colon Cancer Cell Growth and Antioxidant Activity of Bioactive Compounds from Poncirus Trifoliata (L.) Raf. *Bioorg Med Chem.* **2007,** *15* (14), 4923–4932.
57. Lewinsohn, E.; et al. Flavanone Glycoside Biosynthesis in Citrus: Chalcone Synthase, UDP-glucose: Flavanone-7-O-glucosyl-transferase and-rhamnosyl-transferase Activities in Cell-free Extracts. *Plant Phy.* **1989,** *91* (4), 1323–1328.
58. Loizzo, M.; et al. Ruta chalepensis L. (Rutaceae) Leaf Extract: Chemical Composition, Antioxidant and Hypoglicaemic Activities. *Nat. Prod. Res.* **2018,** *32* (5), 521–528.
59. Musiol, R.; Magdziarz, T.; Kurczyk, A. Quinoline Scaffold as a Privileged Substructure in Antimicrobial Drugs. In *Science Against Microbial Pathogens: Communicating Current Research and Technological Advances*; Mendez-Vilas A., Ed.; Formatex, 2011; pp 72–83.

CHAPTER 5

Essential Oils from the Family Rutaceae

AHMED OLATUNDE[1], HABIBU TIJJANI[2], MAYOWA SHAKIRDEEN OBIDOLA[3], NADIA SHARIF[4], UZMA NIHAR[4], NEELMA MUNIR[5], and GODWIN ANYWAR[6]

[1]*Department of Medical Biochemistry, Abubakar Tafawa Balewa University, Bauchi, Nigeria*

[2]*Department of Biochemistry, Bauchi State University, Gadau, Nigeria*

[3]*Department of Crop Production Technology, Federal College of Forestry, Jos, Nigeria*

[4]*Department of Zoology, Women University, Mardan, Pakistan*

[5]*Department of Biotechnology, Lahore College for Women University, Lahore, Pakistan*

[6]*Department of Plant Sciences, Microbiology & Biotechnology, College of Natural Sciences, Makerere University, Kampala, Uganda*

ABSTRACT

Citrus plants belong to the Rutaceae family, and they are edible fruits consumed globally. Members of the *Citrus* family include *Citrus microcarpa*, *Citrus reticulata*, *Citrus paradisi*, *Citrus aurantifolia*, *Citrus hystrix*, *Citrus grandis*, *Citrus sinensis*, *Citrus aurantium*, *Citrus limon*, and *Citrus medica*, among others. The family is rich in active biomolecules such as vitamin C, potassium, flavonoids, coumarins, pectins, folate, and dietary

Phytochemical and Pharmacological Investigation of the Family Rutaceae.
Abdur Rauf, Hafiz Ansar Rasul Suleria, & Syed Muhammad Mukarram Shah (Eds.)

fibers. In addition, these molecules essential oil is an important component of this family of plants. The oil from the *Citrus* species is extracted from the fruits, leaves, peels, and rinds of the plant. *Citrus* plant essential oil can be extracted by employing methods, such as hydrodistillation, steam distillation, and hydrodiffusion, among others. The essential oil contains sesquiterpene, monoterpene hydrocarbons, and other therapeutic molecules. The present chapter highlights the properties of Rutaceae family, and the types and examples of the compounds found in *Citrus* plants as well as some of their therapeutic and non-therapeutic uses.

5.1 INTRODUCTION

Rutaceae (*Citrus*) is one of the main fruit plants produced globally in tropical and subtropical areas and they have high popularity and availability.[31] Because of multiplex biology indicating the presence of apomictic clones and hybrid lines, and the broad geographical availability providing the chance for crossing among species, the taxonomy of Rutaceae is still speculative, indicating a challenge for plant scientists.[14,35] The most predominant fruits include *Citrus reticulate* (tangerine), *C. paradisi* (grapefruit), *C. aurantifolia* (lime), *C. limon* (lemon), and *C. sinensis* (sweet orange). During 2008–2016, the production of *Citrus* fruit was reported to be fairly stable, oscillating between a minimum of 115,541.8 tons in 2009 and a maximum of 132,002.3 tons in 2012 worldwide. In the northern hemisphere, 124,246.0 tons of *Citrus* were recorded this is 97,848.9 tons greater than the ones recorded in the Sothern hemisphere (26,397.1 tons) in 2016. Specifically, the Mediterranean area had 25,216.0 tons and China had 32,705.9 tons, which is more in the boreal hemisphere, while in the Austral region, Brazil (16,555.1 tons) was discovered to have the largest producing house. Furthermore, sweet oranges were reported as the highest produced fruit with 66,974.1 tons, then tangerine which has 32,968.5 tons, followed by limes and lemons with 15,981.8 tons, and grapefruit with 8321.6 tons.[18]

Formerly, the fruits from *Citrus* plant were consumed , mainly as fresh fruits. This is because of extraordinary postharvest stability favoring global trade. However, recently, new economic conditions have been faced and fruit preparation has become a need in order to keep the businesses of farmers going at the local level and to meet the elevating demands of consumers.[7]

Based on this, the processing of canned liqueurs, *Citrus*, jams, marmalades, and essential oil has gained attention and this is associated with small and medium-scale businesses seen globally. In the case of large-scale production, processing started in the early 20th century in Florida and California with industries established mainly for the production of *Citrus* juice.[7]. Most recently (2008–2016), the quantity of fruit to be used for processing has risen from 23,538.9 to 29,400.0 tons, thus showing an elevation from 20% to 25% of the total production of *Citrus* fruits. In the case of sweet orange, it was reported to be the major fruit type that was most processed worldwide with 18,460.9 tons, then limes and lemons with 1821.5 tons, followed by tangerines with 2469.0 tons and grapefruits with 787.5 tons, respectively.[18]

Based on the remarkable number of processed fruits from *Citrus* plant, waste products (about 50%) from the raw *Citrus* fruits were generated.[7,32] In total, waste from *Citrus* can be semisolid and solid residues; this includes fresh peel, membranes, dried peel, membranes, seed, and sludge; the liquid waste which is mainly sludge derived from the liquid waste of the processed *Citrus* and distillery effluents from pectin and citric acid production, molasses, and essential oil plants.[6,59]

The leaves and fruits from *Citrus* plants have different essential oil with several marked flavors and bioactive molecules which are vital to diet and nutrition. These include folic acid, flavonoids, vitamin C, pectin, potassium, coumarins, dietary fibers, and pectins.[47] In many countries, such as Malaysia, it was evidence that oil from *Citrus* species was used as fragrances and flavors, in medical treatment, and in cooking.[40] Furthermore, several compounds have been identified in the essential oil derived from *Citrus* plants. The present chapter provides some *Citrus* plants and the essential oil present in them, extraction methods of these oils, and compounds found in the essential oil as well as some of the therapeutic properties link to them.

5.2 ESSENTIAL OIL FROM CITRUS PLANTS

Essential oil from *Citrus* is broadly used as a raw material in perfume, cosmetic, food, and pharmaceutical industries[48,49,63] and is derived majorly from the fruit rind and other parts including the leaves and flowers. Their semi-volatile and volatile composition are made up of about 85–99% of the total oil contents from the *Citrus* plants,[15,20,57,63] containing more than 200 identified compounds. Hydrocarbon and compounds such as monoquiterpenes and sesquiterpenes are with the highest report, then olefinic and aliphatic C6–C12 non-terpene esters, aldehydes, ketones, acids, and alcohols together with many other aromatic molecules. The identified non-volatile residues include flavonoids, diterpenoids, coumarins, fatty acids, and sterols.[20] Furthermore, several volatile and semivolatile phytocompounds were stated to be present in the essential oil obtained from flowers, leaves, and rinds of *Citrus* plants. The essential oil includes *C. reticulate* (mandarin), *C. aurantium* (bitter orange), *C. limon* (lemon), *C. grandis* (pummel), *C. medica* (citron), *C. sinensis* (sweet orange), and *Citrus bergamia* (bergamot orange). Their oil compositions have shown several medicinal benefits (Table 5.1).

5.2.1 C. MICROCARPA

C. microcarpa has branches with many spines, twigs, and stems. The leaves of this plant are between 2and 3-cm thick and 2.5–6.8-cm long. The plant is used in the management of cough, fever, and pharyngitis.[45] Traditionally, the juice is used for alleviating respiratory disorders, growth stimulant, and bone strengthening. In addition, fruits are usually used in cooking as additives and flavoring agents. The leaves are used in the management of skin disorders, and treat headache and sore throat.[37,48] The essential oil extracted from *C. microcarpa* peels was indicated to have β-myrcene (1.8%), limonene (94.0%), linalool (0.4%), and α-terpineol (0.3%). In the leaves of the plant, sesquiterpene hydrocarbons are found in high concentration. They include hedycaryol (19.0%), α-sesquiphellandrene (18.3%), and β-eudesmol (8.6%). Essential oil from *C. microcarpa* is obtained by hydrodistillation for 8 h.[22,47]

5.2.2 C. AURANTIFOLIA

The plant is mostly known as common lime and it is used in traditional medicine and cuisine.[46] It has a spiny stem and is approximately 3–5-m tall. *C. aurantifolia* has leaves that are ovate-shaped and are 3–5 cm thick and 5–9 cm long. The fruits from this plant are green and change to yellow after ripening and the flowers are white.[55] *C. aurantifolia* essential oil has different sesquiterpene and monoterpene hydrocarbons, including limonene, which is the most abundant compound.[46] In traditional settings, the plant is used for decreasing blood fat, sugar, and cholesterol and to ease the digestion process.[56] The oil obtained from the fruits of the plants is reported to be used to alleviate arthritis, cold, bronchitis, and asthma.[26] Furthermore, juice from the fruit is employed for preventing pimples, managing bleeding of the uterus, serving as a poison antidote, and elevating stamina.[2, 24] *C. aurantifolia* fruit juice is reported to be a remarkable cough alleviator when mixed to honey and sugar and can decrease body temperature, serve as a meat softener and remove body smell.[55] The plant is also essential as a repellent for moths, mosquitoes, and cats and it exhibits anti-inflammatory and antioxidant actions.[16,30,56]

C. aurantifolia essential oil from peels was evidenced to contain terpinen-4-ol (2.0%), geranial (2.1%), α-terpineol (2.4%), neral (5.3%), geraniol (7.5%), β-pinene (28.4%), and limonene (39.3%). Essential oil from the leaves of the plant contains β-caryophyllene (5.7%), geranyl acetate (6.6%), geraniol (7.5%), nerol (9.5%), neral (11.4%), limonene (16.4%) and geranial (19.4%).[22]

5.2.3 C. HYSTRIX

C. hystrix is also called wild lime and it contains leaves and fruits. The leaves of the plant are 7.5–10-cm long and the height of the whole *C. hystrix* is approximately 3–5 m. About 4–6 petals are found on its white flowers and the pear-shaped fruits have a diameter of approximately 5–7.5 cm with dark green coloration and yellow when ripe.[4, 47] *C. hystrix* essential oil is utilized in aromatherapy, and in industries involving beauty products and cosmetics.[65] Traditionally, the plant is used for managing hypertension, infant diarrhea, fever, abdominal pain, and flu.[19] The fruits are employed as a blood purifier, digestive stimulant, flavoring agent, for decreasing high blood pressure, and used in the production of shampoo for washing hair.[25] Juice from the fruit is used for removing body odor and softening of the skin.[13] Several pharmacological actions such as antileukemic, antioxidant, antitussive, and antibacterial properties are linked to the essential oil.[44]

Reports indicated that *C. hystrix* essential oil obtained from Malaysia mainly contains β-pinene, citronellal, limonene, and sabinene.

Furthermore, hydrodiffusion steam distillation, steam distillation, and automated steam distillation process are the basic procedures for treating essential oil obtained from *C. hystrix* peel.[23,34,38] In another study, in the same country, it was reported that the essential oil extracted from the fresh leaves of the aforementioned plant contains limonene (5–7%), β-citronellol (10–14%), and citronellal (61–73%) and this oil was obtained by Likens–Nikerson extraction and steam distillation methods.[43] In another study, involving the extraction of oil from *C. hystrix* peel by water distillation method, terpinen-4-ol (8.9%), citronellal (11.9%), limonene (14.2%), and β-pinene (39.3%) were isolated as the major phytochemicals whereas the extraction of essential oil from the leaves of the plant by the water distillation methods shows the presence of citronellyl acetate (4.1%), β-citronellol (6.7%), citronellal (72.4%), limonene (0.1%), and β-pinene (1.9%).[22] In addition, essential oil from *C. hystrix* is evidenced to show marked antibacterial action against *Escherichia coli*, *Staphylococcus aureus*, and *Bacillus subtilis*.[40] More so, essential oil from the plant exhibited insecticidal actions against *Spodoptera litura* and it shows repellent activity against the larvae of the insect following 24- and 48-h treatments.[27]

5.2.4 C. GRANDIS

C. grandis is *Citrus* plant found in countries like India, Thailand, Bangladesh, Vietnam, Philippines, and Malaysia. The fruit has a thickness of about 10–30 cm, and height approximately 5–15 m with leaves of 2–12-cm thick and 5–20-cm long. The fruits have a pear shape with 10–30-cm width, and greenish yellow or pale yellow color,[37,47,53] Also, the plant has therapeutic properties including a cure for fever, arthritis, ulcers, gout, and kidney disease.[46] The peels and fruit pulp are employed as a stomach tonic and appetizer and also for managing cough and inflammation and the juice from the fruit promotes weight loss and influences a decrease in cholesterol levels. Also, the fruits of the plant are used in the cosmetic, pharmaceutical, food, and perfume industries.[47,61] The essential oils extracted from the fruits and leaves of *C. grandis* are utilized as a part of different toiletry products and its pungent flowers are used in perfume industries.[63] Using steam distillation for the extraction of peel essential oil from the aforementioned plant, compounds such as *cis*-carveol (1.5%), β-myrcene (2.2%), and limonene (81.6%) were reported.[53] However, the Likens–Nikerson procedure results to *cis*-carveol (1.4%), β-myrcene (1.6%), and limonene (86.8%) from the essential oil peel of *C. grandis*. Essential oils obtained from the leaves of the plant gave sesquiterpene and monoterpene hydrocarbons as the main compounds present. These include δ-3-carene (3.9%), geranial (4.5%), β-pinene (4.9%), *trans*-β-ocimene (9.9%), α-cadinene (7.1%), β-caryophyllene (15.4%), and phytol (23.1%).[22] The solvent extraction and head–space solid-phase microextraction methods are used for the extraction of essential oils from the peels of pink and white *C. grandis*. In another study, extracted essential oil from the blossoms of white *C. grandis* by head–space solid-phase microextraction method was shown to contain linalool (9.2%), *cis*-β-ocimene (12.0%), and limonene (48.2%) whereas from the pink variety of the plant from blossoms contains *cis*-β-ocimene (4.0%), limonene (15.5%), and linalool (56.5%).[10,47]

5.2.5 C. RETICULATA

C. reticulata are another type of plant from the *Citrus* family. The essential oils extracted from this plant seem to be different from other species as a result of the presence of certain distinctive non-terpenoid aldehyde phytocompounds including (*Z*)-2-dodecenal and (*E,E*)-2,4-heptadienal. Also many other non-terpenoid aldehyde phytocompounds have been majorly extracted in the *C. reticulata* peel and this includes (*E*)-2-nonenal, (Z)-4-decenal, (*E*)-2-octenal, (2*E*,4Z,7*Z*)-decatrienal, (*E,Z*)-2,4-nonadienal, (*Z*)-5-dodecenal, and (*Z*)-6-dodecenal.[8,12,39,51,54] In the rinds of the plant, sesquiterpene hydrocarbons such as β-copaene have been documented.[29] Moreover, monoterpene hydrocarbons are the most predominant compounds found in the essential oils extracted from *C. reticulata*. These compounds includes β-citronellal (0.6%), linalool (2.9%), β-pinene (4%), α-pinene (0.1–3.93%), β-myrcene (0.1–7.43%), γ-terpinene (15%), and limonene (60–95%).[17,36,50,60] Other compounds found in the essential oils of the plant are in the range of 0.1–0.7% including the aromatic phytocompound (methyl *N*-methylanthranilate), non-terpene aliphatic phytocompounds (decanal and octanal), and sesquiterpene (α-sinensal) and the non-terpenoids aldehydes, which do not typically exceed 0.1%.[20]

TABLE 5.1 Plant Names, Essential Oil Composition, and Associated Medical Benefits.

Plant name	Common names	Essential oil	Associated medicinal benefits	References
C. microcarpa	Calamansi	Limonene (94.0%), linalool (0.4%), β-myrcene (1.8%), and α-terpineol (0.3%)	Alleviating respiratory disorders, growth stimulants, bone-strengthening cough, fever, and pharyngitis	[45]
C. aurantifolia	Common lime or key lime	Terpinen-4-ol (2.0%), geranial (2.1%), α-terpineol (2.4%), neral (5.3%), geraniol (7.5%), β-pinene (28.4%), limonene (39.3%), β-caryophyllene (5.7%), geranyl acetate (6.6%), geraniol (7.5%), nerol (9.5%), neral (11.4%), limonene (16.4%), and geranial (19.4%)	Arthritis, cold, bronchitis, and asthma	[22, 47]
C. hystrix	Wild lime or kaffir lime	Limonene (14.2%), β-citronellol (10–14%), citronellal (61–73%), terpinen-4-ol (8.9%), citronellal (11.9%), β-pinene (39.3%), citronellyl acetate (4.1%), β-citronellol (6.7%), and citronellal (72.4%)	Antileukemic, antioxidant, antitussive, antibacterial, antihypertension, antidiarrhea, fever, abdominal pains, flu, and insecticidal	[19, 22, 43, 44]
C. grandis	Pomelo	*cis*-Carveol (1.5%), δ-3-carene (3.9%), β-myrcene (2.2%), limonene (81.6%), *cis*-carveol (1.4%), β-myrcene (1.6%), limonene (86.8%), geranial (4.5%), β-pinene (4.9%), *trans*-β-ocimene (9.9%), α-cadinene (7.1%), β-caryophyllene (15.4%), and phytol (23.1%), linalool (9.2%); *cis*-β-ocimene (12.0%) and limonene (48.2%); *cis*-β-ocimene (4.0%), limonene (15.5%), and linalool (56.5%)	Fever, arthritis, ulcers, gout, and kidney disease	[22, 46, 53]
C. reticulata	Mandarin orange	β-Citronellal (0.6%), linalool (2.9%), β-pinene (4%), α-pinene (0.1–3.93%), β-myrcene (0.1–7.43%), γ-terpinene (15%), and limonene (60–95%)	Snakebite, fever, edema, stomachache, and cardiac *diseases*	[36, 17, 50, 60]
C. sinensis	Sweet orange	Limonene, 2-butyl-2-octenal, 2-octyl-2-dodecenal, and 2-hexyl-2-decenal	Organoleptic properties, angina, anxiety, bronchitis, cold, colic, constipation, cough, cramps, depression, diarrhea, hypertension, menstrual disorder, obesity, stress, and tuberculosis	[1, 20, 33, 41, 58]

TABLE 5.1 *(Continued)*

Plant name	Common names	Essential oil	Associated medicinal benefits	References
C. paradisi	Grapefruit	γ-Terpinene, linalool, nootkatone, and undecanoic acid	Anti-inflammatory	[42, 50]
C. aurantium	Bitter orange	β-Pinene, limonene, octanal, β-myrcene, (*E*)-β-ocimene, linalool, geranyl acetate (0.9%), and linalyl acetate (5.0%) α-ocimene, α-terpinen-4-ol acetate	Malarial fever, anti-inflammation, fever, and diarrhea	[5, 28, 50, 51]
C. limon	Lemon	(*E*)-α-Bisabolene, (Z)-α-bisabolene, 2-(2-methyltetrahydrothiophen-2-yl) ethyl acetate, 2-(5-isopropyl-2-methyltetrahydrothiophen-2-yl) ethyl acetate, 2-(5-isopropyl-2-methyltetrahydrothiophen-2-yl) ethanol, 2-(5-isopropylidene-2-methyltetrahydrothiophen-2-yl) ethyl acetate, 2-[5-(1-hydroxy-1-methylethyl)2-methyltetrahydrothiophen-2-yl] ethyl acetate, 2-propanoylthiophene, 3-mercapto-3,7-dimethyl-6-octenyl acetate, 3-methyldodecanal, 3-methylundecanal, 4-methyldodecanal, 4-methyltridecanal, 4-methylundecanal, 6-methyldecanal, campherenol, sulfur non-terpenoid, α-bisabolol, β-santalene, and γ-curcumene	Antibacterial, antidepression, and anti-inflammatory	[9, 20, 50]
C. medica	Citron	α-Cuprenene, italicene, 9-epi-caryophyllene, γ-cuprenene, longifolene, β-oplopenone, nootkatol, 9-epi-caryophyllene, longifolene and *trans*-4-caranone, camphene (10%), γ-terpinene (31%), and β-pinene (9.7%)	Analgesic, anthelmintic, antibacterial, anticancerous, anti-catarrhal, antidiabetic, antifungal, antihyperglycemic, anti-hypertensive, antimicrobial, antiulcer, capillary protector, cardioprotective, diuretic, estrogenic, and antioxidant	[3, 11, 20]

5.2.6 C. SINENSIS

The compounds found in the essential oils extracted from *C. sinensis* is the most studied among other *Citrus* species. Also, the plant was reported to be higher in several types of sesquiterpene hydrocarbons than in other Citrus species. Examples of these compounds present in *C. sinensis* include aromadendrene and sesquiphellandrene. Monoterpenoids (such as limonene) are another type of compounds reported to be found in large amounts in the essential oils of the plant.[41,58] In a recent study, compounds such as 2-octyl-2-dodecenal and 2-hexyl-2-decenal, and 2-butyl-2-octenal are the non-terpene aldehydes found in the folded cold press orange oil at small concentration.[1] The aforementioned phytocompounds are decanal, octanal, and hexanal self-aldol condensation products, respectively. The products possess attractive organoleptic properties according to taste and aroma assessments.[20]

5.2.7 C. PARADISI

C. paradisi has compounds present in its essential oil comparable to those of *C. sinensis* but lower compounds in to *C. reticulata* and *C. grandis*.[20] Also, several sesquiterpene hydrocarbons in the plant have been documented. The main compound reported in *C. paradisi* includes γ-terpinene, linalool, nootkatone, undecanoic acid, and others.[42,50]

5.2.8 C. AURANTIUM

Monoterpenoids such as eucalyptol, geranyl propionate, and isomenthol are the numerous compounds found in *C. aurantium*.[21] The compounds found in the plant are similar to those found in *C. reticulata* and *C. sinensis* (β-pinene, (*E*)-β-ocimene, limonene, octanal, β-myrcene, and linalool) although in *C. aurantium*, the percentage of linalyl acetate and geranyl acetate is approximately 5.0% and 0.9%, respectively[28,50] whereas in *C. reticulata* and *C. sinensis* the percentage of these phytocompounds are under 0.1%.[36,41] Moreover, it was reported that the peel of *C. aurantium* contains 0.87% of eucalyptol, and this is higher in other *Citrus* species. Similarly, the plant contains the highest level of germacrene D (sesquiterpene) than the other species.[28] Other compounds found in *C. aurantium* are mainly the monoterpene hydrocarbons such as α-ocimene, and oxygenated phytocompounds including α-terpinen-4-ol acetate.[5,52]

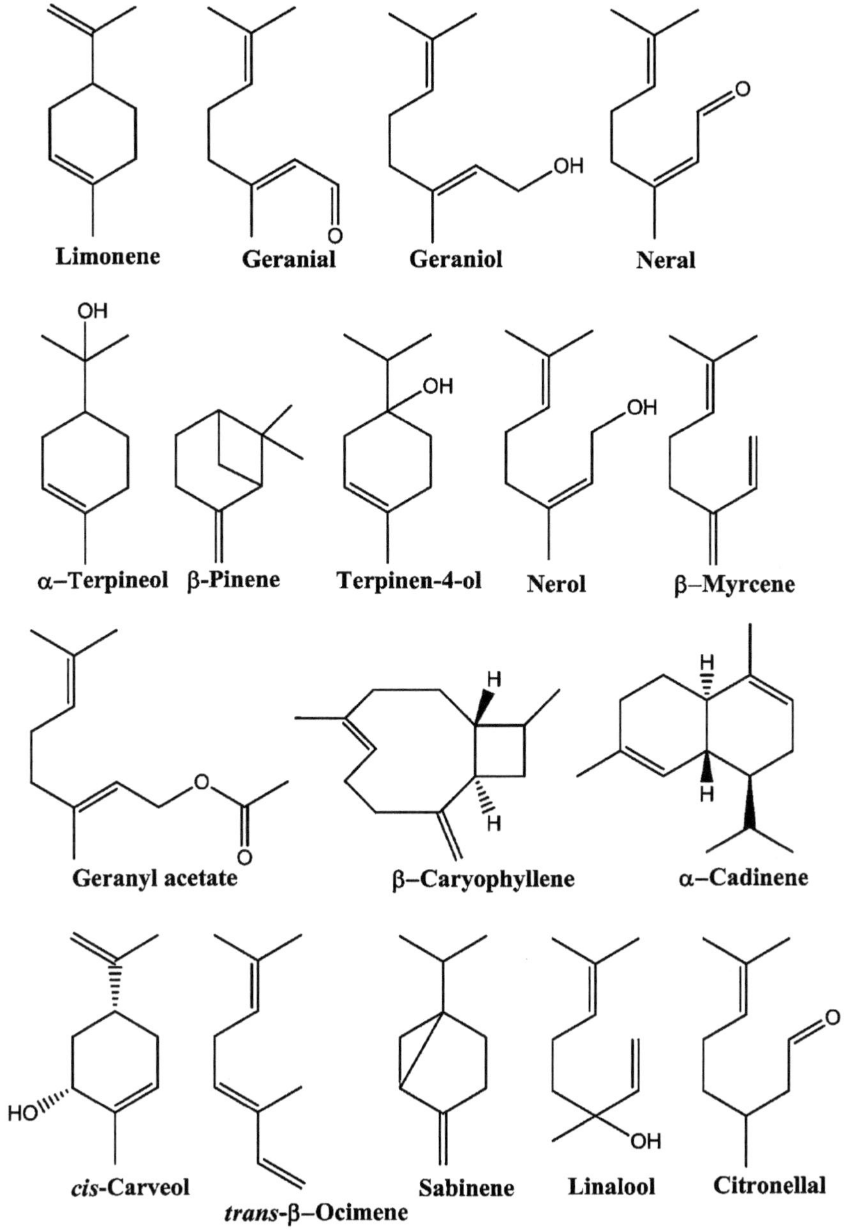

FIGURE 5.1 Some of the compounds found in the essential oil derived from *Citrus* species.

5.2.9 C. LIMON

Essential oil from lemon is the most different from other *Citrus* plants based on volatiles. Most of the compounds found in the plant are similar to those of the species stated earlier.[20] However, more than 150 compounds were found in the peel of *C. limon* by dichloromethane extraction and they were not identified in other species of *Citrus*.[9,20] These compounds include 3-mercapto-3,7-dimethyl-6-octenyl acetate, 2-(5-isopropyl-2-methyltetrahydrothiophen-2-yl)ethanol, 2-(5-isopropylidene-2-methyltetrahydrothiophen-2-yl) ethyl acetate, 2-(5-isopropyl-2-methyltetrahydrothiophen-2-yl) ethyl acetate, 2-[5-(1-hydroxy-1-methylethyl)2-methyltetrahydrothiophen-2-yl] ethyl acetate and sulfur non-terpenoid compounds, 2-(2-methyltetrahydrothiophen-2-yl) ethyl acetate, 3-methylundecanal, 3-methyldodecanal, 2-propanoylthiophene. The branched-chain aliphatic compounds (6-methyldecanal), and aldehydes such as 4-methyldodecanal, 4-methylundecanal, and 4-methyltridecanal are present in the plant as traces.[20]

Based on quantitative analysis, the main compound seen in the essential oil from *C. limon* is limonene at a range of 48–70%, and this is similar to the phytocompound found in *C. aurantifolia*. The next predominant phytocompounds are the terpenoids and they are similar to those found in *C. aurntifolia*.[20] Also, certain types of sesquiterpene compounds were found in the peel of *C. limon* this includes *(E)*-α-bisabolene and α-bisabolol, *(Z)*-α-bisabolene, γ-curcumene, β-santalene, and campherenol.[9,50] It was stated that the profiling of the majority of these sesquiterpenes might not be completely dependable as their standards are not commercially available.[20] Another compounds found in *C. limon* and other species is norcarotenoid 6-methyl-5-hepten-2-one.[9,50]

5.2.10 C. MEDICA

C. medica is a member of the *Citrus* family and it has compounds similar to other members. A study indicated that the plant contains some compounds which are rare in other *Citrus* species, they are *cis*-4-caranone or dehydrosabina ketone, monoterpenes cuminyl alcohol, and particularly sesquiterpene such as α-cuprenene, italicene, 9-epi-caryophyllene, γ-cuprenene, longifolene, β-oplopenone, nootkatol, and others. However, it was reported the majority of these compounds were tentatively identified due to lack of standards for their identification with some exceptions in the case of 9-epi-caryophyllene and longifolene and *trans*-4-caranone.[3,20] The most predominant compound in this specie of *Citrus* is limonene[64] whereas other monoterpenes such as β-pinene, γ-terpinene, and camphene are seen at a higher level compared to *C. reticulata*. In the oil from *C. medica*, the camphene, γ-terpinene, and β-pinene have percentage of 10, 31, and 9.7%, respectively. Furthermore, the plant shows a higher level of germacrene D and (*E*)-α-bergamotene.[3,50] The peel of *C. medica* does not contain certain phytocompounds, particularly the non-terpenoid aldehyde but they are present in other species of *Citrus*. In addition to these, compounds including (*E*)-2-dodecenal, (*E,E*)-2,4-decadienal, (*E*)-2-hexenal, (*Z*)-3-hexenal, α-sinensal, and carvacrol have not been documented in *C. medica*.[20]

5.3 CONCLUSIONS

The essential oil extracted from the peels, fruits, and leaves of *Citrus* plant can be obtained by various methods including steam distillation, hydrodiffusion, and head–space solid-phase microextraction. This oil is used in cosmetics, perfumes, pharmaceutics, and food industries due to its fragrance and flavors as well as for cooking. *Citrus* species essential oil contains sesquiterpene and monoterpene hydrocarbons, with limonene documented as the most predominant example among other compounds. Furthermore, the essential oil exhibited non-therapeutic actions such as insect repellent and showed remarkable bioactivities such as antimicrobial, antioxidant, and anti-inflammatory properties.

KEYWORDS

- **essential oil**
- **Rutaceae**
- **extraction methods**
- **biological activity**
- ***Citrus* plant**

REFERENCES

1. Abreu, I.; da Costa, N. C.; van Es, A.; Kim, J.-A.; Parasar, U.; Poulsen, M. L. Natural Occurrence of Aldol Condensation Products in Valencia Orange Oil. *J. Food Sci.* **2017,** *82*, 2805–2815. DOI: 10.1111/1750-3841.13948.
2. Aibinu, I.; Adenipekun, T.; Adelowotan, T.; Ogunsanya, T. & Odugbemi, T. Evaluation of the Antimicrobial Properties of Different Parts of *Citrus aurantifolia* (Lime Fruit) as Used Locally. *Afr. J. Tradit. Complemnet. Altern. Med.* **2007,** *4*, 185–190.
3. Aliberti, L.; Caputo, L.; de Feo, V.; de Martino, L.; Nazzaro, F.; Souza, L. F. Chemical Composition and In Vitro Antimicrobial, Cytotoxic, and Central Nervous System Activities of the Essential Oils of *Citrus medica* L. cv. 'Liscia' and *C. medica* cv. 'Rugosa' Cultivated in Southern Italy. *Molecules* **2016,** *21*, E1244. DOI: 10.3390/molecules21091244.
4. Awang, C. R. C. & Minyak, K. Esensial *Citrus* spp Ke Atas Sistem Kardiovaskular Tikus Serta Kesan Antiresah dan Antidepresi Pada Mencit. (Anti-Anxious and Anti-Depression Effects of *Citrus* spp. Essential Oils Towards Cardiovascular System of Mice). Master's Thesis, Universiti Sains Malaysia, Pulau Pinang, Malaysia, 1 May 2007.
5. Baik, J. S.; Kim, S. S.; Lee, J. A.; Oh, T. H.; Kim, J. Y.; Lee, N. H. et al. Chemical Composition and Biological Activities of Essential Oils Extracted from Korean Endemic *Citrus* Species. *J. Microbiol. Biotechnol.* **2008,** *18*, 74–79.
6. Bampidis, V. A.; Robinson, P. H. *Citrus* By-products as Ruminant Feeds: A review. *Anim. Feed Sci. Technol.* **2006,** *128* (3–4), 175–217.
7. Berk, Z. *Citrus Fruit Processing*; Academic Press: Cambridge, 2016.
8. Buettner, A.; Mestres, M.; Fischer, A.; Guasch, J.; Schieberle, P. Evaluation of the Most Odour-Active Compounds in the Peel Oil of Clementines (*Citrus reticulata* Blanco cv. Clementine). *Eur. Food Res. Technol.* **2003,** *216*, 11–14. DOI: 10.1007/s00217-002-0586-y.
9. Cannon, R. J.; Kazimierski, A.; Curto, N. L.; Li, J.; Trinnaman, L.; Janczuk, A. J. et al. Identification, Synthesis, and Characterization of Novel Sulfurcontaining Volatile Compounds from the In-depth Analysis of Lisbon Lemon Peels (*Citrus limon* L. Burm. f. cv. Lisbon). *J. Agric. Food Chem.* **2015,** *63*, 1915–1931. DOI: 10.1021/jf505177r.
10. Cheong, M. W.; Loke, X. Q.; Liu, S. Q.; Pramudya, K.; Curran, P.; Yu, B. Characterization of Volatile Compounds and Aroma Profiles of Malaysian Pomelo (*Citrus grandis* (L.) Osbeck) Blossom and Peel. *J. Essent. Oil Res.* **2011,** *23*, 34–44.
11. Chhikara, N.; Kour, R.; Jaglan, S.; Gupta, P.; Gat, Y.; Panghal, A. *Citrus medica*: Nutritional, Phytochemical Composition and Health Benefits—A Review. *Food Funct.* **2018,** *9* (2), 1978-1992. DOI: 10.1039/c7fo02035j
12. Chisholm, M. G.; Jell, J. A.; Cass, D. M. J. Characterization of the Major Odorants Found in the Peel Oil of *Citrus reticulata* Blanco cv. Clementine Using Gas Chromatography-Olfactometry. *Flavour Frag. J.* **2003,** *18*, 275–281.
13. Dassanayake, M. D. *A Revised Handbook to the Flora of Ceylon*, Vol. V; Amerind Publishing Co Ltd.: New Delhi, India, 1985; pp 432–433.

14. de Arau'jo, E. F.; de Queiroz, L. P.; Machado, M. A. What Is *Citrus*? Taxonomic Implications from a Study of cp-DNA Evolution in the Tribe Citreae (Rutaceae subfamily Aurantioideae). *Organ. Divers. Evol.* **2003,** *3* (1), 55–62.
15. Dugo, P.; Mondello, L. *Citrus Oils: Composition, Advanced Analytical Techniques, Contaminants, and Biological Activity*, Vol. 49; CRC Press: Boca Raton, 2011.
16. Effiom, O. E.; Avoaja, D. A.; Ohaeri, C. C. Mosquito Repellent Activity of Phytochemical Extracts from Fruit Peels of *Citrus* Fruit Species. *Glob. J. Sci. Front. Res.* **2012,** *12*, 5–8.
17. Fanciullino, A. L.; Tomi, F.; Luro, F.; Desjobert, J. M.; Casanova, J. Chemical Variability of Peel and Leaf Oils of Mandarins. *Flavour Frag. J.* **2006,** *21*, 359–367. DOI: 10.1002/ffj.1658.
18. FAO. *Citrus* Fruit Fresh and Processed: Statistical Bulletin **2016,** 2017. http://www.fao.org/3/a-i8092e.pdf (accessed 14 Feb 2019).
19. Fortin, H.; Vigora, C.; Lohezic-Le, F.; Robina, V.; le Bosse, B.; Boustiea, J.; Arnoros, M. In Vitro Antiviral Activity of Thirty-Six Plants from La Reunion Island. *Fitoterapia* **2002,** *73*, 346–350.
20. González-Mas M. C.; Rambla, J. L.; López-Gresa, M. P.; Blázquez, M. A.; Granell, A. Volatile Compounds in *Citrus* Essential Oils: A Comprehensive Review. *Front. Plant Sci.* **2019,** 10–12. DOI: 10.3389/fpls.2019.00012
21. Jabri Karoui, I.; Marzouk, B. Characterization of Bioactive Compounds in Tunisian Bitter Orange (*Citrus aurantium* L.) Peel and Juice and Determination of Their Antioxidant Activities. *Biomed. Res. Int.* **2013,** 345415. DOI: 10.1155/2013/345415.
22. Jantan, I.; Ahmad, A. S.; Ahmad, A. R.; Ali, N. A. M.; Ayop, N. Chemical Composition of Some *Citrlus* Oils from Malaysia. *J. Essent. Oil Res.* **1996,** *8*, 627–632.
23. Kasuan, N.; Muhammad, Z.; Yusoff, Z.; Rahiman, M. H. F.; Taib, M. N.; Haiyee, Z. A. Extraction of *Citrus hystrix* DC (Kaffir Lime) Essential Oil Using Automated Steam Distillation Process: Analysis of Volatile Compounds. *Malays. J. Anal. Sci.* **2013,** *17*, 359–369.
24. Khare, C. P. *Indian Medicinal Plants: An Illustrated Dictionary*; Springer: Berlin, Germany, 2007; p 154.
25. Koh, D.; Ong, C. N. Phytophotodermatitis Due to the Application of *Citrus hystrix* as a Folk Remedy. *Br. J. Dermatol.* **1999,** *140*, 737–738.
26. Kunow, M. A. *Maya Medicine: Traditional Healing in Yucatan*; University of New Mexico Press: Albuquerque, 2003; p 117.
27. Loh, F. S.; Awang, R. M.; Omar, D.; Rahmani, M. Insecticidal Properties of *Citrus hystrix* DC Leaves Essential Oil Against *Spodoptera litura* Fabricius. *J. Med. Plants Res.* **2011,** *5*, 3739–3744.
28. Lota, M. L.; de Rocca Serra, D.; Jacquemond, C.; Tomi, F.; Casanova, J. Chemical Variability of Peel and Leaf Essential Oils of Sour Orange. *Flavour Fragrance J.* **2001,** *16*, 89–96.
29. Luciardi, M. C.; Blázquez, M. A.; Cartagena, E.; Bardón, A.; Arena, M. E. Mandarin Essential Oils Inhibit Quorum Sensing and Virulence Factors of *Pseudomonas aeruginosa. LWT Food Sci. Technol.* **2016,** *68*, 373–380. DOI: 10.1016/j.lwt.2015.12.056.
30. Lyle, S. *How to Use Citrus Fruit Peels in the Home and Garden: In Discovering Fruit and Nuts*; David Bateman Ltd.: Auckland, 2006; pp 130–142.

31. Malacrida, C. R.; Kimura, M.; Jorge, N. Phytochemicals and Antioxidant Activity of *Citrus* Seed Oils. *Food Sci. Technol. Res.* **2012,** *18* (3), 399–404.
32. Marın, F. R.; Soler-Rivas, C.; Benavente-Garcıa, O.; Castillo, J.; Perez-Alvarez, J. A. By-Products from Different *Citrus* Processes as a Source of Customized Functional Fibers. *Food Chem.* **2007,** *100,* 736–741.
33. Milind P.; Chaturvede D. Orange: Range of Benefits. *Int. Res. J. Pharm.* **2018,** *3,* 59–63.
34. Mohd-Yusoff, Z. Muhammad, Z.; Kasuan, N.; Rahiman, M. H. F.; Taib, M. N. (2013). Effect of Temperature on Kaffir Lime Oil by Using Hydro-Diffusion Steam Distillation System. *Malays. J. Anal. Sci.* **2013,** *17,* 326–339.
35. Moore, G. A. Orange and Lemons: Clues to the Taxonomy of *Citrus* from Molecular Markers. *Trends Genet.* **2001,** *17* (9), 536–540. DOI: 10.1016/S0168-9525(01)02442-8.
36. Mondello, L.; Casilli, A.; Tranchida, P. Q.; Cicero, L.; Dugo, P.; Dugo, G. (2003). Comparison of Fast and Conventional GC Analysis for *Citrus* Essential Oils. *J. Agric. Food Chem.* **2003,** *51,* 5602–5606. DOI: 10.1021/jf0341971.
37. Morton, J. F. Mexican Lime. In *Fruits of Warm Climates,* 1st ed.; Creative Resource Systems: Winterville, 1987; pp 168–172.
38. Muhammad, Z.; Mohd Yusoff, Z.; Nordin, M. N. N.; Kasuan, N.; Taib, M. N.; Rahiman, M. H. F.; Haiyee, Z. A. Steam Distillation with Induction Heating System: Analysis of Kaffir Lime Oil Compound and Production Yield at Various Temperatures. *Malays. J. Anal. Sci.* **2013,** *17,* 340–347.
39. Naef, R.; Velluz, A. (2001). Volatile Constituents in Extracts of Mandarin and Tangerine Peel. *J. Essent. Oil Res.* **2001,** *13,* 154–157. DOI: 10.1080/10412905.2001.9699647.
40. Ng, D. S. H.; Rose, L. C.; Suhaimi, H.; Mohammad, H.; Rozaini, M. Z. H.; Taib, M. Preliminary Evaluation on the Antibacterial Activities of *Citrus hystrix* Oil Emulsions Stabilized by Tween 80 and Span 80. *Int. J. Pharm. Pharm. Sci.* **2011,** *3,* 209–211.
41. Njoroge, S. M.; Koaze, H.; Karanja, P. N.; Sawamura, M. Essential Oil Constituents of Three Varieties of Kenyan Sweet Oranges (*Citrus sinensis*). *Flavour Frag. J.* **2005a,** *20,* 80–85.
42. Njoroge, S. M.; Koaze, H.; Karanja, P. N.; Sawamura, M. Volatile Constituents of Redblush Grapefruit (*Citrus paradisi*) and Pummelo (*Citrus grandis*) Peel Essential Oils from Kenya. *J. Agric. Food Chem.* 2005b, *53,* 9790–9794. DOI: 10.1021/jf051373s.
43. Nor, O. M. Volatile Aroma Compounds in *Citrus hystrix* Oil. *J. Trop. Agric. Food Sci.* **1999,** *27,* 225–229.
44. Norkaew, O.; Pitija, K.; Pripdeevech, P.; Sookwong, P.; Wongpornchai, S. Supercritical Fluid Extraction and Gas Chromatographic-Mass Spectrometric Analysis of Terpenoids in Fresh Kaffir Lime Leaf Oil. *Chiang Mai J. Sci.* **2013,** *40,* 240–247.
45. Ong, H. C. *Buah: Khasiat Makanan & Ubatan; Utusan*; Publications & Distributors: Kuala Lumpur, 2004; p 90.
46. Orwa, C.; Mutua, A.; Kindt, R.; Jamnadass, R.; Simons, A. Agroforestree Database: A Tree Reference and Selection Guide Version 4.0, 2009http://www.worldagroforestry.org/af/treedb/

47. Othman, S. N. A. M.; Hassan, M. A.; Nahar, L.; Basar, N.; Jamil, S.; Sarker, S. D. Essential Oils from the Malaysian *Citrus* (Rutaceae) Medicinal Plants. *Medicines* **2016,** *3*, 13; DOI: 10.3390/medicines3020013.
48. Othman, S. N. A. M.; Hassan, M. A.; Nahar, L.; Basar, N.; Jamil, S.; Sarker, S. D. Essential Oils from the Malaysian *Citrus* (Rutaceae) Medicinal Plants. *Medicines* **2017,** *3*, 13. DOI: 10.3390/medicines3020013.
49. Palazzolo, E.; Laudicina, V. A.; Germanà, M. A. (2013). Current and Potential Use of *Citrus* Essential Oils. *Curr. Org. Chem.* **2013,** *17*, 3042–3049. DOI: 10.2174/13852728113179990122.
50. Petretto, G. L.; Sarais, G.; Maldini, M. T.; Foddai, M.; Tirillini, B.; Rourke, J. P. et al. (2016). *Citrus* monstruosa Discrimination Among Several *Citrus* Species by Multivariate Analysis of Volatiles: A Metabolomic Approach. *J. Food Process. Preserv.* **2016,** *40*, 950–957. DOI: 10.1111/jfpp.12674.
51. Pino, J. A.; Quijano-Celís, C. E. Chromatographic Deterpenation of Mandarin Cold-Pressed Oil. *J. Essent. Oil Bear. Plant* **2007,** *10*, 504–509. DOI: 10.1080/0972060X.2007.10643587.
52. Pino, J. A.; Rosado, A. Composition of Cold-Pressed Bitter Orange Oil from Cuba. *J. Essent. Oil Res.* **2000,** *12*, 675–676. DOI: 10.1080/10412905.2000.9712187.
53. Roger, G. D. P. *Encyclopedia of Medicinal Plants Education and Health*, 1st ed.; Library editorial safeliz SL: Malaga, 2002; pp 153–154.
54. Ruberto, G.; Rapisarda, P. Essential Oils of New Pigmented *Citrus* Hybrids: *Citrus sinensis* L. Osbeck × C. clementina Hort. ex Tanaka. *J. Food Sci.* **2002,** *67*, 2778–2780. DOI: 10.1111/j.1365-2621.2002.tb08815.x.
55. Saidan, I. *Dalam Dusun Melayu (in Malay Orchard)*; Dewan Bahasa dan Pustaka: Kuala Lumpur, 2013; pp 243–244.
56. Samah, B. Serangan, J.; Punca, P.; Kaedah, M. *Heart Attack: The Cause, Prevention & Treatment*; Alaf 21: Selangor, 2009; pp 104–105.
57. Sarrou, E.; Chatzopoulou, P.; Dimassi-Theriou, K.; Therios, I. Volatile Constituents and Antioxidant Activity of Peel, Flowers and Leaf Oils of *Citrus aurantium* L. Growing in Greece. *Molecules* **2013,** *18*, 10639–10647. DOI: 10.3390/molecules180910639.
58. Sawamura, M.; Minh Tu, N. T.; Yu, X.; Xu, B. Volatile Constituents of the Peel Oils of Several Sweet Oranges in China. *J. Essent. Oil Res.* **2005a,** *17*, 2–6. DOI: 10.1080/10412905.2005.9698813.
59. Sharma, K.; Mahato, N.; Cho, M. H.; Lee, Y. R. Converting Citrus Wastes Into Value-Added Products: Economic and Environmently Friendly Approaches. *Nutrition* **2017,** *34*, 29–46.
60. Tao, N.; Jia, L.; Zhou, H. Anti-Fungal Activity of *Citrus reticulata* Blanco Essential Oil Against *Penicillium italicum* and *Penicillium digitatum*. *Food Chem.* **2014,** *153*, 265–271. DOI: 10.1016/j.foodchem.2013.12.070.
61. Thavanapong, N.; Wetwitayaklung, P.; Charoenteeraboon, J. Comparison of Essential Oils Compositions of *Citrus maxima* Merr. Peel Obtained by Cold Press and Vacuum Stream Distillation Methods and of Its Peel and Flower Extract Obtained by Supercritical Carbon Dioxide Extraction Method and Their Antimicrobial Activity. *J. Essent. Oil Res.* **2010,** *22*, 71–77.
62. Tomiyama, K.; Aoki, H.; Oikawa, T.; Sakurai, K.; Kashara, Y.; Kawakami, Y. Characteristic Volatile Components of Japanese Sour *Citrus* Fruits: Yuzu, Sudachi and Kabosu. *Flavour Frag. J.* **2012,** *27*, 341–355. DOI: 10.1002/ffj. 3104.

63. Tranchida, P. Q.; Bonaccorsi, I.; Dugo, P.; Mondello, L.; Dugo, G. Analysis of *Citrus* Essential Oils: State of the Art and Future Perspectives: A Review. *Flavour Frag. J.* **2012,** 27, 98–123. DOI: 10.1002/ffj.2089.
64. Verzera, A.; Trozzi, A.; Zappalá, M.; Condurso, C.; Cotroneo, A. Essential Oil Composition of *Citrus meyerii* Y. Tan. and *Citrus medica* L. cv. Diamante and Their Lemon Hybrids. *J. Agric. Food Chem.* **2005,** *53*, 4890–4894. DOI: 10.1021/jf047879c.
65. Waikedre, J.; Dugay, A.; Barrachina, I.; Herrenknecht, C.; Cabalion, P.; Fournet, A. Chemical Composition and Antimicrobial Activity of the Essential Oils from New Caledonian *Citrus macroptera* and *Citrus hystrix*. *Chem. Biodivers.* **2010,** *7*, 871–877.

CHAPTER 6

The Family Rutaceae: A Comprehensive Review of Its Phytochemical and Pharmacological Perspectives

KISHWAR SULTANA[1], AISHMA KHATTAK[2], MUHAMMAD RIZWAN[3], ZAHID HUSSAIN[3], ABDUR RAUF[4], and MOHAMMED A. AL-DUAIS[5,6]

[1]*Centre of Biotechnology & Microbiology, University of Peshawar, KPK, Pakistan*

[2]*Department of Bioinformatics, Shaheed Benazir Bhutto Women University Peshawar, KP, Pakistan*

[3]*Centre for Biotechnology and Microbiology, University of Swat, KP, Pakistan*

[4]*Department of Chemistry, University of Swabi, Swabi, Anbar, Khyber Pakhtunkhwa, KP, Pakistan*

[5]*Department of Biochemistry, Faculty of Sciences, University of Tabuk, Tabuk, Saudi Arabia*

[6]*Biochemistry Unit, Chemistry, Department, Faculty of Sciences, IBB University, IBB, Yemen*

ABSTRACT

The current review discusses the traditional information and pharmacological features of the most important medicinal species that are the focus of this study. The selected species for the present review belongs to the

Phytochemical and Pharmacological Investigation of the Family Rutaceae.
Abdur Rauf, Hafiz Ansar Rasul Suleria, & Syed Muhammad Mukarram Shah (Eds.)

family Rutaceae which comprises the most medicinally important sources of secondary metabolites, mainly such as flavonoids, alkaloids, limonoids, coumarins, essential oils, and volatile oils. The main purpose of this review study is to look in to the more recent developments in the phytochemical and pharmacological investigation of this family as well as the traditional uses of its species in a systematic way. The species belongs to this family known for largest numbers of secondary metabolites. Other species in this family have, however, been the subject of pharmacological investigations. The species comprises a wide range of biologically active chemical compounds such as antibacterials, anticoagulants, and anti-carcinogens in it. Accordingly, anti-inflammatory, antihyperlipidemic, antidepressant, antibacterial, antiviral, anthelmintic, antifungal, insecticidal, antiprotozoal, cytotoxic, and antioxidant effects of the species of family Rutaceae considered here. This current paper provides an overview of its traditional, phytochemical, and pharmacological features, which may aid researchers in determining the potency and efficacy of Family Rutaceae as therapeutic plants.

6.1 INTRODUCTION

The rue family Rutaceae consist of 160 genera and approximately 2070 species of wooded shrubs as well as trees and it is widely dispersed in tropical areas and warm moderate areas[1] The Rutaceae family members have wide application in traditional medicine, cuisine, perfumery and numerous publications about the occurrence of secondary metabolites have been reported.[2] The survey of phytochemical studies of this family revealed the presence of volatile oils, flavonoids, alkaloids, coumarins, and limonoids[3] and they possess diverse biological activities such as antimicrobial,[4] antidiarrhoeal,[5] anticholinesterasic,[6] antileishmanial,[7] antiprotozoal,[8] larvicidal[9] as well as antioxidant properties.[10]

In the Rutaceae family, there are seven subfamilies, one of them is *Aurantioideae.*[11] *Micromelum* is the solitary genus of the *Micromelinae* subtribe and *Clauseneae* is one of the two tribes of subfamily *Aurantioideae*. The genus *Micromelum* has yielded alkaloids, flavonoids, and coumarins.[12] The species of genus *Micromelum* contains little spineless like trees and locally the specie *Micromelum minutum* is identified in Malayasia as "Cherek-cherek," "Chemomar," "Kematu," or "Secherek".[13,14] The genus

Micromelum species has exposed a wide range of biologically active constituents such as antibacterial, anticarcinogenic as well as anticoagulating agent[15–17] in it.

Rutaceae, generally recognized as a family of *citrus* which is a flowery plant family with around one hundred and sixty genera also having species of flowers. In the family, the most significant and economical genera are *Citrus* which comprises the *Citrus limon* (lemon), *Citrus aurantifolia* (mostly the lime), *Citrus paradisi* (grapefruit), *Citrus sinensis* (orange) as well as *Agathosma* and *Zanthoxylum* or *Fagara*.

6.2 PHARMACOLOGICAL INVESTIGATION OF FAMILY RUTACEAE

6.2.1 *FAGARA LEPRIEURII*

Fagara leprieurii belongs to the genus *Fagara* and family Rutaceae. The genus *Fagara* species have several antimicrobial properties and has traditionally utilized to treat cancer, kidney ache, ulcer, gonorrhea, gingivitis, gastritis, bilharzia, sterility, diarrhea, laxative as well as other various pathogenic diseases.[18,19]

In vitro biological assays revealed that the essential oil of *F. leprieurii*, methanolic extracts and some of the compounds isolated from fruits of the plant exhibited moderate antimicrobial as well as anticancer activities.

6.2.2 *FRAXINUS XANTHOXYLOIDES*

Fraxinus xanthoxyloides belongs to the family Rutaceae and have traditionally use as antiseptic and laxative.[19] Ngane et al. have assessed the antifungal assay of the 90% of watery-ethanol extracts obtained from foliage, stem barks as well as roots of *F. leprieurii* in addition to the *F. xanthoxyloides* contrary to nine fungal isolates. They discovered that these extracts, to erratic extents, repressed the in vitro growth of *C. neoformans* and *C. albicans* in addition to some filamentous fungi such as *Trichophyton rubrum*, *Trichophyton mentagrophytes*, *Microsporum gypseum*, *B. cinerea*, *Scopulariopsis brevicaulis*, *Aspergillus fumigatus*, and *A. flavus*. They also noticed that just the extracts acquired from the *F. xanthoxyloides* stem bark and roots had antifungal action on the considered microorganisms, with changing value of MIC from 0.125 to 1 mg/mL for the stem bark and from 0.5 to 1 mg/mL for the roots.

6.2.3 *MICROMELUM MINUTUM*

Micromelum minutum belonging to the family Rutaceae has traditionally been utilized to treat ague and ringworm, as well as to treat fever and regulate menstruation.[17,20] It is also used to treat tumors in traditional medicine of Thai.[21] The *M. minutum* stems are traditionally utilized as carminative, leaves as febrifuge, and the fruits in addition to flowers, respectively, as a purgative and an expectorant.[22] Hemorrhoids, bad breath, teething disorders in babies, white scum on tongue, and toothache are all treated using the juice extracted from the leaves. The leaves of *M. minutum* are used as a general tonic, and the shoots of *M. minutum* are utilized to cure infantile convulsions. A poultice made from the pulverized leaves is utilized to alleviate skin allergy. Scabies irritants are relieved by rubbing the leaves over the skin. The inner bark of the twigs and leaves are utilized to treat a variety of ailments, including stomach-ache and headache, sore tongue and coughs, heavy menstruation, thrush and gonorrhea. Headache is treated with the bark's fluid, while stomachache is treated with bark infusion.[23] The roots of *M. minutum* are utilized as a febrifuge and an infusions or decoction are used as a carminative and to treat diarrhea in children. They're said to help with toothache and can also be utilized to cure headache as well as stomachache. Coughs are treated with root pieces of *M. minutum* by chewing with betel. This plant is utilized to treat headache.[23]

M. minutum has been found to have cytotoxic properties against a variety of cell lines. The chloroform extracts of *M. minutum leaves exhibited maximum effect.* The extracts of *M. minutum bark* displayed moderate action against a T-lymphoblastic leukemia (CEM-SS) cell line,[24] with IC_{50} values of 13.7 and 4.2 μg/mL, respectively, indicating that they have great potential to be used as anti-cancer drug.[24]

The action of mahanine, a main ingredient of the Thai vegetable *M. minutum* edible sections, by activating the pathway of apoptosis in U937 human cells of leukemia reported by Roy et al. earlier.[25] According to the findings, mahanine compound induced pathway of apoptosis in U937 human cells via mitochondrial pathway and inhibited growth of cell.

Several end points were employed to check for apoptotic pathway in this study, including DNA analysis, morphological alterations occur in cells, phosphatidylserine membrane bound translocation and the relative numbers of apoptotic as well as viable cells. The activation of caspases, particularly caspase-3, loss of membrane permeability of mitochondria, reduced ATP levels in the cell and cytochrome *c* release in the cytosol were discovered to be involved in apoptosis of mahanine-induced in U937 human cells. Because mitochondrial membrane permeability is identified as to regulate the releasing of cytochrome *c*, the findings suggested that mahanine was primarily targeting mitochondria. So, according to the study, the loss of mitochondrial membranes permeability caused due to mahanine, which leads to cause apoptosis and activation of caspase-3. For apoptosis, the IC_{50} value of mahanine was calculated as 8.7 μM after 12 h.[25]

The colorimetric procedure was used to perform a cytotoxic activity against cell line such as cholangiocarcinoma (KKU-100), as reported by Skehan and his co-workers earlier.[26] The compounds isolated form *M. minutum fruits including* micromelin, microminutin, minumicrolin murralonginol, minumicrolin, murralongin, and scopoletin, all showed cytotoxic effect against the cell line cholangiocarcinoma (KKU-100). Murrangatin and microminutin, in particular, exhibited maximum cytotoxic effect against the cell line cholangiocacinoma (KKU-100) having IC_{50} values of 2.9 and 1.7 μg/mL, respectively.[27] Other isolated compounds had IC_{50} values were calculated as micromelin (9.2 μg/mL), minumicrolin (10.2 μg/mL), murralonginol (10.0 μg/mL), murralongin (9.0 μg/mL), and scopoletin (19.2 μg/mL).[27]

The chloroform extract of *M. minutum* leaves was used for isolation of compounds and two compounds such as 8-hydroxyisocapnolactone-2', 3'-diol and 2', 3'-epoxyisocapnolactone were identified. Both the compounds were used for evaluating cytotoxicity against cell lines HL60. The compounds have significantly showed cytotoxic effect against HL60 cell lines having the IC_{50} value of 8-hydroxyisocapnolactone-2', 3'-diol was 2.5 μg/mL and 2', 3'-epoxyisocapnolactone was 4.2 μg/mL, respectively, against the cancer cell line.[28]

The compound such as 8-hydroxyisocapnolactone-2. 3-diol obtained from the chloroform extract of *M. minutum leaves demonstrated maximum* cytotoxic effect against HeLa cell line of cervical cancer having respective IC_{50} value was 6.9 μg/mL.[28]

The compound such as 8-hydroxyisocapnolactone-2, 3-diol obtained from the chloroform extract of *M. minutum leaves exhibited highest* cytotoxic effect against the HepG2 cell line of liver cancer with maximum IC_{50} was 5.9 µg/mL.[28]

The MTT colorimetric cell viability activity was used to determine the cytotoxic assay against A549 and SBC3 cell lines of lung adenocarcinoma.[29] The compounds such as 8-hydroxyisocapnolactone-2, 3-diol, minutin B in addition to Clauslactone E obtained from the *M. minutum leaves exhibited maximum* cytotoxic action against A549 and SBC3 cell lines of lung adenocarcinoma with estimated IC_{50} values for 8-hydroxyisocapnolactone-2, 3-diol (10.1 µM), minutin B (17.5 µM), and clauslactone E (3.7, 10.4 µM), respectively.[29]

The MTT colorimetric cell viability activity was used to evaluate the cytotoxic assay against K562/ADM as well as cell lines K562 of leukemia. Compounds such as 8-hydroxyisocapnolactone-2, 3-diol, minutin B and Clauslactone E obtained from *M. minutum leaves exhibited highest* cytotoxic activation against K562/ADM and cell lines K562 of leukemia having IC_{50} values were calculated for 8-hydroxyisocapnolactone-2, 3-diol (16.9 and 10.1 µM), minutin B (8.7 and 6.7 µM), and clauslactone E (12.1 and 10.8 µM), respectively.[29]

A compound recognized as Phebalosin, derived from *M. minutum* stem bark, inhibited the crown gall tumors growth on potato discs by 60% as well as 70%, in two separate tests.[30]

A compound known as Murralonginol, derived from *M. minutum* fruits for the first time, was showed mild cytotoxic action against KB cell line of human epidermoid having IC_{50} value for murralonginol isovalerate (41.1 and 30.4 µg/mL) and 7-demethylmurralonginol isovalerate also showed mild cytotoxic action with IC_{50} value calculated as (17.8 µg/mL) respectively.[31]

Three compounds obtained from *M. minutum fruits including* murralonginol isovalerate, murralonginol, and 7-demethylmurralonginol isovalerate were showed mild cytotoxic action against the cell line having the IC_{50} values of 49.5, 27.1, and 46.5 µg/mL, respectively.[31]

The crude extract of *M. minutum* leaves *was assessed for* leishmanicidal action, and it was discovered that the methanolic extract showed 68% mortality against Leishmania major at 100 µg/mL concentration. Four compounds such as clauslactone E, 8-hydroxyisocapnolactone-2',3'-diol, minutin B, and minutin A were derived from *M. minutum* leaves exhibited noteworthy cytotoxic action against L. major having IC_{50} values calculated as 9.8, 12.1, 20.2, and 26.2 μM, respectively.[31]

6.2.4 ZANTHOXYLI FRUCTUS

Zanthoxyli fructus belongs to the family Rutaceae which relieves pain (abdominal pain), expels cold sensation in the abdomen and known for antiparasitic, mildly diuretic, analgesic, antibiotic, and antidiarrhea. For weaning purposes combine with brown sugar or take as a capsule (grind the herb, insert in capsules and take 400 g three times daily for 3–4 days); also used as a cooking ingredient.

6.2.5 RUTA GRAVEOLENS

Ruta graveolens is a member of Rutaceae and is an important medicinal plant widely utilized in the Mediterranean area for the treatment of ache, dermatitis, rheumatism, and further diseases of inflammation, but because of its possible toxicity its usage is restricted. Nevertheless, various new topographies have been characterized for this therapeutic plant, including its shielding influence opposing to mutagenesis and breakage of DNA strand. Additionally, rue has been revealed to persuade the elimination of group of amide from the antiapoptotic protein Bcl-xL in malignant cells of human brain, but not in typical T and B lymphocytes, therefore, rise the susceptibility of cancer cells to death, whereas leave normal cells remain unaffected.

6.2.6 PHELLODENDRON AMURENSE

Phellodendron amurense which belongs to family Rutaceae *which used as* antibiotic, antihypertensive, antitussive, expectorant, and antiasthmatic, used to treat chronic bronchitis; used in treatment of meningitis, diarrhea, and conjunctivitis (particularly in children). The topical application of it for trichomonal cervicitis, vaginitis, eczema on the ear. It comprises a compound berberine belongs to alkaloids group.

6.2.7 *MURRAYA PANICULATE*

Murraya is a genus of floral plants, closely associated with the *Citrus* genus. It is in the *Clauseninae* subtribe, which are renowned precisely as the secluded citroid produce trees. *Murraya paniculata* (L.) Jack, commonly recognized as Orange Jessamine, is a tropical, perennial plant with small, snowy, aromatic flowers, which is cultivated as a decorative tree or hedgerow. It is associated with the Rutaceae family and can be generally found in Australia and South Asia. In traditional medicine, several parts of this plant have been castoff. In Bangladesh, extracts of the leaves of *M. paniculata* is used orally to lessen pain.[32] In the Philippines, these leaves were used to cure dysentery and diarrhea due to their stimulating and caustic actions.[33] In India, individuals occasionally used *M. paniculata* root bark as medicine for panic, coughs, and rheumatism.[34] Moreover, broiled leaves and boiled brushwood useful for lessen swollen joints and stomach pain, respectively, in India.[35] There are numerous information on pharmacological effects of the plant encompassing antinociceptive,[32,36] antioxidant[37,38], and antidiabetic,[35] to antimicrobial[35] and pain-relieving activities.[39] Various research groups have described isolation of effective constituents like alkaloids,[40] phenols,[41] terpenoids[42], and flavonoids[35,43,44] from fruits, leaves, floras, and barks of root of *M. paniculata* as healthiness treatment. In prospect of the significance of *M. paniculata*, it is identified for antioxidant, antidiabetic, antinociceptive, antimicrobial, and pain-relieving properties.

6.2.8 *AEGLE MARMELOS*

Aegle marmelos is a species of wood apple belongs to the Rutaceae family that can be found in China, India, Nepal, Ceylon, Sri Lanka, Pakistan, Myanmar, Bangladesh, Vietnam, Cambodia, Laos, Indonesia, Thailand, Sri Lanka, Tibet, Malaysia, Fiji, Java, and Philippines.[45]

A. marmelos has traditionally been used to cure a variety of ailments. Traditionally, it is utilized for the treatment of dysentery, jaundice, constipation, stomachic, stomachache, ulcers, fever, inflammations, chronic diarrhea, asthma, upper respiratory tract infections, acute bronchitis, febrile delirium, acidity, snakebite, leucoderma, thyroid disorders, abdominal discomfort, epilepsy, spermatorrhoea, burning sensation, leporsy, indigestion, smallpox, myalgia, eye disorders, swelling, nausea, mental illnesses, sores, thirst, and tumors.[45]

In the Rutaceae family, *A. marmelos* specie is one of the most commonly used nutraceuticals as well as medicinal plant. The plant has been linked to a variety of medicinal benefits in recent years.

Antioxidizing agents are the substances that have free radicals scavenging potential and can protect cells from oxidative damage caused by free radicals. Natural sources, like plants, can create antioxidant chemicals. The occurrence of isocatechins, catechins, flavonoids, flavones, isoflavones, coumarin lignans and anthocyanin in these plants possess antioxidant action. *A. marmelos* is said to have antioxidizing properties against a variety of free radicals. The fruit of *A. marmelos* has been shown to have antioxidant properties. The free radical scavenging potential and antioxidant action of *A. marmelos* ready and underdeveloped fruit were found to be linked. When compared to unripe fruit extract, the enzymatic antioxidants found in ripe fruit were shown to be higher except glutathione peroxidase. In immature fruit, the proportion of free radical inhibition was likewise higher than in mature fruit.[45]

A. marmelos is commonly used to treat a variety of infectious disorders and has been described as having the ability to halt a wide variety of diseases causing by microorganisms. Many in vitro investigations have confirmed studies confirmed *A. marmelos* extracts as antimicrobial potential against pathogens such as bacteria and fungi which causing diseases.

The antimicrobial properties of *A. marmelos* leaves extract were evaluated with the help of agar well diffusion method. The antimicrobial action of the aqueous, petroleum ether, and ethanolic extract obtained from *A. marmelos* leaves against *Salmonella typhi, Klebsiella pneumoniae*, *Streptococcus pneumoniae*, *Proteus vulgaris*, and *Escherichia coli* was demonstrated. The ethanolic extract was shown to be effective against *Penicillium chrysogenum* while the petroleum ether and aqueous extracts were found to be effective against *Fusarium oxysporum.*[46]

The antidiarrheal action is one of the *A. marmelos* most important therapeutic properties and it has long been used to treat dysentery as well as chronic diarrhea. Several in vivo and in vitro experiments have recently been undertaken to confirm *A. marmelos* antidiarrheal properties.

The fleshy tissue of *A. marmelos* dried fruit was found to have antidiarrheal property in vitro. The MIC technique was used to achieve antidiarrheal action against diarrhea causing pathogenic organisms. The

A. marmelos ethanolic extract was found to have virulent action against *S. flexneri, S. sonnei*, and *Shigella boydii* as well as a mild effect on *S. dysenteriae*.[47]

The aqueous extracts obtained from leaves of *A. marmelos* were reported to check antifertility effect in Albino male rats. The rats were given *A. marmelos* aqueous extracts obtained from leaves in a concentration of 250 mg/kg weight of body for 45 days. The weights of the epididymis, seminal vesicle as well as testis were reduced as a result of the treatment. Epididymal sperm count and motility, testicular sperm count as well as irregular sperm count were all reduced by the extract.[48]

A. marmelos is widely utilized in traditional medicine, and its fruits are commonly consumed as a source of nutrition. However, *A. marmelos* should not be used by pregnant or breastfeeding women because *A. marmelos* leaves have been used to sterilize women as well as to induce abortion in the past.[45]

6.2.9 *ZANTHOXYLUM ARMATUM*

Zanthoxylum armatum is one of the most important medicinal plant, often recognized as timur, is a member of Rutaceae family. The plant is extensively distributed in Pakistan to eastward China, Korea, Japan, and Northern India.[49]

The fruits, seeds as well as bark are extensively utilized as carminative, anthelmintic, and stomachic in traditional medicine. In preliminary testing, the stem showed hypoglycemic activity. The bark has a sour taste and is used to clean teeth. In terms of temperature and dyspepsia, the fruits and seeds act as a perfumed tonic. The fruits extract *Z. armatum* was found to be active in excreting roundworms from the stomach of children. Due to its deodorant, antiseptic and disinfecting characteristics, the fruits are utilized in dental problems and in scabies lotion. Deodorant, antiseptic, and disinfecting characteristics are also found in the *Z. armatum* essential oil.

The essential oil obtained from *Z. armatum* seeds was utilized to repel three species of mosquito such as *Anopheles stephensi*, *Aedes aegypti*, and *Culex quinquefasciatus*. The essential oil research found a minimum of 28 compounds, comprising largely of oxygenated monoterpenes. The three mosquito species larvae were vulnerable to the composition of aromatic oil.[50]

The ethanolic extract of *Z. armatum* obtained from stem bark showed antioxidant activities. In vivo, antioxidant action was judged in Wistar species of rats by the use of Carrageenan induced paw edema, whereas in vitro assay was achieved by utilizing DPPH free radical technique. The antioxidant activity of plant extract displayed momentous effect.[51]

The effect of an ethanolic extract of *Z. armatum* leaves on CCl4-induced hepatotoxicity in rats was demonstrated by a considerable reduction in liver enzymes and inflammation as well as histopathological examination of the liver. The hepatoprotective action results were shown to be noteworthy.[52]

The essential oil extracted from *Z. armatum* had a large and rapid poisoning effect on *Culex quinquefasciatus* and *Aedes albopictus*, shows a potential as natural pesticides against mosquitoes.[53]

The result of *Z. armatum* crude extract elicited concentration-dependent relaxation of spontaneous contractions as well as increase K+ influenced contractions in isolated rabbit's jejunum. The findings indicated that *Z. armatum* spasmolytic actions are caused by a Ca++ antagonistic mechanism, which gives a pharmacologic basis for its medical usage in respiratory, gastrointestinal as well as cardiovascular problems.

The coumarin (Bergapten) extracted from *Zanthoxylum* significantly reduces the production of two pro-inflammatory cytokines, interleukin-6 (IL 6) as well as tumor necrosis factor-α (TNF-α). Linalyl acetate and Linalool, two further compounds, have been found to have anti-inflammatory properties.[54]

6.2.10 GLYCOSMIS PENTAPHYLLA

Glycosmis pentaphylla Corr. is a member of the Rutaceae family. The leaves and stems of this tiny shrub are bitter.[55]

The isolated compounds such as Skimmianine and Arborine from *G. pentaphylla* demonstrated considerable antibacterial efficacy against clinically isolated MDR *S. aureus* strains with inhibition of growth zones ranging from 25 to 28 mm. Skimmianine and Arborine have a dose-dependent bactericidal action, according to the data. 101 MDR *S. aureus* in addition to 410 strain showed high susceptibility to the obtained compounds, having highest zone of inhibition growth and lowest values of MIC (0.2 μg/mL) as well as MBC (0.2 μg/mL). Likewise, the isolated compound skimmianine exhibited the maximum antibacterial action against 315 strain of MDR *S. aureus* with inhibition of growth zones ranging from (28 mm) having lowermost values of MIC (0.2 μg/mL) as well as MBC (0.2 μg/mL). The antibacterial action was observed in lower concentration having a greater effect of inhibition. The isolated compounds such as Skimmianine and Arborine were found to have a high growth inhibitory action on all pathogens examined. Overall, the Skimmianine and Arborine were found to have higher antibacterial efficacy than conventional medicines. The isolated compounds such as Skimmianine and Arborine with low MIC and MBC concentrations had a surprising level of bactericidal action against the MDR *S. aureus* strains examined.

Antibacterial efficacy of the isolated Skimmianine and Arborine compounds showed noteworthy action against MDR *S. aureus* isolates. Bowen et al.,[56] Chakravarti et al.,[57] and Chakravarty et al.[58] reported the same results and found that the compounds like Skimmianine and Arborine have antibacterial action against both Gram-positive bacterium (*S. aureus as well as* MRSA) which were found noneffective against the strain *S. aureus* and Gram-negative bacterium (*E. coli as well as S. typhimurium*). Jeyachandran et al.[59] reported earlier that the isolation of the compound plumbagin extracted from the *Plumbago zeylanica root*, which is a bioactive component and has a higher potential of toxicity against the strain *S. aureus*.

The time kill kinetic activity was used to analyze the bacterial mechanism studies of compounds. This assay was used to evaluate the bacterial viability after treatment and determine the minimal time required to provide the bactericidal efficacy. The bactericidal action of both compounds such as Skimmianine and Arborine exhibited similar time-killing kinetic patterns against the strains of MDR *S. aureus*. The bactericidal action of both Skimmianine and Arborine compounds against MDR *S. aureus* strains was steadily increased by the time up to 12 h of exposure at their MBC respective concentrations for both the strains and within this time period, the multiple drug-resistant *S. aureus* strains were found killed. As indicated in time kill curves relative to control cells, both the Skimmianine and Arborine compounds have a time-dependent and exhibited swift bactericidal action against the strains of MDR *S. aureus*, resulting in bacterial mortality at the initial phase of stationary. The time-kill test determines changes in antibacterial activity rate and extent over time and can also give information about growth kinetics.[60] The overall outcome of this assay demonstrated that both Skimmianine and Arborine compounds had highest bactericidal action against *S. aureus* as compared to the medium in laboratory.

The Skimmianine and Arborine compounds isolated from *Glycosmis pentaphylla* have been shown to promote protein leakage by enhancing the permeability of membrane in multiple drug-resistant strains *S. aureus*. In order to check the effect of compounds such as Skimmianine and Arborine alone on protein leakage in all strains, the cells showed 55% protein leakage by the treatment of skimmianine compound for 100 μg/mL, resulting in 55% protein leakage and with the treatment of 75 μg/mL arborine compound, resulting in 54% protein leakage in all strains. The quantity of protein produced from the cells was increased with the treatment of cells with compounds such as Skimmianine and Arborine compared with the control such as commercial standard antibiotics with recommended doses. The overall evaluation of Skimmianine and Arborine compounds for protein leakage assay exhibited noteworthy effect against the tested MDR *S. aureus* strains. Similar findings were observed regarding to protein leakage assay when medicinal plant oil components were used to treat patients which supports the current work.[61]

The isolated compounds like Skimmianine and Arborine were further examined for their potential efficacy on surfaces of cell to give knowledge better about the mechanism of antibacterial action. The changes occur in morphology of all strains of MDR *S. aureus* as well as MTCC-96 *S. aureus* strain the reference standard were evaluated by using SEM analysis after treatment of Skimmianine and Arborine compounds having antibacterial efficacy. The untreated bacterial cells (control strains) in MDR *S. aureus* visualized as cocci (grape) shaped and the cell surfaces were found intact with no damage. The cocci (grape-shaped) properties of *S. aureus* cells were lost after the treatment with Skimmianine and Arborine compounds. Furthermore, uneven fragments were detected, indicating that the cell membrane of bacteria had been damaged. Likewise, the treated cell surfaces of *S. aureus* were observed uneven, the cell size was condensed in addition to the cells appeared injured. The current study used the SEM analysis to confirm that the membrane surfaces of MDR *S. aureus* were degraded after treating with the obtained Skimmianine and Arborine compounds from *G. pentaphylla*. The current study of SEM analysis results demonstrated that the action of Skimmianine and Arborine compounds has a considerable impact on cell membrane of bacteria. The bioactive compounds such as

Skimmianine and Arborine have been linked to the disruption of bacterial cell membrane and cell death. Similarly, Campos et al.[62] reported the same results, who discovered that the cytoplasm and membranes of *E. coli* and *S. aureus* were extremely impacted by the passive diffusion of plant metabolites into the cells, resulting cell disintegration. These differences occurring in bacterial cells could be induced by membrane lysis in addition to alteration triggered by the degradation of membrane permeability as well as integrity caused by Skimmianine and Arborine isolated compounds. As a result, the differences may cause harm to the cell's internal materials.[63,64] The results of SEM analysis were consistent with those discoveries of Paul et al.[65] and Sharma et al.[66] stated that further antimicrobial was used to treat cells. The damaged cell membrane of *S. aureus* reduces the outflow of cellular components as well as cell permeability, leading to cell death, according to Shen et al.[67] which validated the current investigation.

6.2.11 CITRUS

The family name for genus *Citrus* is Rutaceae. It falls under the broad category of trees and bushes. The subtropical as well as tropical areas of the world are most suitable for the growth of this genus like tangerines, lemons in addition to oranges.[68]

Citrus fruits are extensively used for the treatment of different diseases. The incidence of gastric cancer could be reduced by the intake of citrus fruits.[69] The central nervous system of human beings is affected by some compounds derived from the fruits of these citrus species. Some recent studies have found that limonene could be useful as an effective anxiolytic agent for both animals and human beings. Limonene is present in citrus aurantium in a higher concentration.[70,71]

Apart from the above-mentioned advantages, some valuable nutrients like dietary fibers, pectin, coumarins, flavonoids, potassium, vitamin C, vitamin B9 as well as essential oils could be extracted from other parts of the plants (leaves, flowers, fruits, and peels).[72] In addition, citrus essential oils (CEOs) are explored as antidiabetic, antifungal, antibacterial, and antioxidant. They have found their application in cosmetic, medical, sanitary, agricultural, as well as food industries.[73,74] They have been widely employed in medicines, textiles, cosmetics in addition to formulations of food.[75] Furthermore, due to the refreshing aroma, CEOs are also effective disinfectants and vacuum fresheners. The chances of Alzheimer's disease could be narrowed down by making a pharmaceutical formulation using compounds of *Citrus limon* (lemon) in addition to β-cyclodextrin of *Citrus sinensis* (derived from orange EO).[76]

The negative effects of various food components on human health have become a major concern for food industries and consumers. As an alternative to nonorganic compounds in foods, organic and natural compounds are gaining immense importance because of being cost effective, little or no side effects, beneficial to health, and environment friendly. The commercially synthesized chemical preservatives for maintaining the food quality and safety could be effectively replaced by plants derived natural antimycotics.[77–79] Out of all the available plant essential oils, CEOs are of center of interest. This is because of their high yield, aromas, and flavors and widely applied insecticidal, antibacterial, and antifungal properties.[80–83] Additionally, CEOs have found their application in preparations, packaging, as well as preservation of food in order to strengthen the food control.[84] This review will provide an insight into the different methods of extraction, purification, and detection of CEOs obtained from different species of citrus, their components and their widespread application in food industry.

A survey was carried out for *in vitro* as well as *in vivo* studies to understand the therapeutic potential of different species of *Citrus*. The results are briefly given in Table 6.1.

6.2.12 PHARMACOLOGICAL ACTIONS OF CITRUS SPECIES

6.2.12.1 CITRUS AURANTIUM

C. aurantium is inherent plant species of Southeastern Asia, including China and India. This specie is known by a popular name *laranjeira cavalo* or *laranjeira amarga* (bitter orange). Arabs introduced this plant species to Syria and Egypt and later brought to Europe.[85] It was widely employed as a general tonic, sedative, digestive, tranquilizer, antidote against poisons, appetite, cardiac, and vascular stimulant in the Mediterranean region during medieval times.[86] Traditionally, this plant is utilized as a general tonic and gastrointestinal in Chinese medicine system.[87]

In Brazilian communities, the ethnopharmacological studies documented the *C. aurantium* peels, leaves, flowers, and fruit have been used to cure disorders like sleeplessness, hysteria as well as anxiety.[88, 89] The preparation of tea from *C. aurantium* leaves is reported to combat stomach acidity, stomach cramps, constipation, and fever whereas diabetes is controlled by the tea prepared from seeds.[90]

The effect of *C. aurantium* on the nervous system was investigated, particularly its anxiolytic actions. The anxiolytic effect has been verified through numerous experiments on both humans as well as animals. The anxiolytic action was shown by the essential oil derived from *C. aurantium* peels after administering a single dose to rats which signified an increment in the dwelling time of open arms of the elevated plus maze test.[91] Furthermore, the anxiolytic action was confirmed in experimental models of obsessive compulsive as well as generalized anxiety disorder. Even after treatment of 15 consecutive days, the mice did not reflect any symptoms of motor impairment.

In a research work reported by Leite et al., two models of anxiety were selected such as open field maze as well as elevated plus maze tests. For this study, 12 rats were inhaled with *C. aurantium* essential oil at 1.0, 2.5, and 5.0% concentrations for duration of 7 min enclosed in a box made of acrylic. A possible central action was confirmed by reduction in the emotional reaction in both models of animal.[92] The effect of *C. aurantium* on nervous system was explained by a proposed mechanism in which 5-HT1 as a receptor and serotonin receptors as a subtype.[93]

The anxiolytic effect of *C. aurantium* L. was further confirmed through clinical trials. In this study, the scale of STAI was utilized to evaluate the levels of anxiety. The subject was given distilled flower extract of *C. aurantium* for 2 h prior to the process. It was found that the experimental group showed an improvement in the anxiety level when compared to control group.[94] In another reported work, the process of medullary material collection was applied to patients with chronic myelogenous leukemia. It was observed that the patients inhaled with essential oils of *C. aurantium* L. showed lower anxiety levels during the procedure. Even a single dose exhibited a comparable action to that of the anxiolytic employed as a positive control. This further verifies its usefulness in the reduction of patient anxiety.

In another study, the anxiolytic effect was tested for primiparous women (18–35 years old) were exposed to *C. aurantium* aromatherapy at the time of labor. A significant decrease in the anxiety level was observed for women under observation. The anxiolytic action prior to labor was also verified. In another clinical trial, 126 primiparous women were kept under observation during the first stage of labor. Each woman was inoculated with a dose of 4 mL *C. aurantium* soaked in distilled water after every 30 min. Women under observation showed pain relief during first stage of labor. About 88.1% of the participants were satisfied with the treatment procedure and 92.1% showed positive response to use this process in the future. This is a simple, cost effective, and noninvasive interaction which could be proved beneficial for pregnant women.[95] However, further studies are still required in this direction, as the birth from vagina is a serious and painful problem and women are commonly feared about labor.[96]

Apart from its actions on nervous system, research has been performed to investigate additional reliable *C. aurantium* effects. The anti-inflammatory effect of flavonoids like hesperidin, naringin and nobiletin derived from *C. aurantium* was checked on cells of rat *in vitro*. The anti-inflammatory effect of flavonoids was confirmed by the suppression of pro-inflammatory mediators.[97] The anticancer effect of flavonoids was also evaluated in murine lung cells. The anticancer activity was confirmed through the cell migration and regulation of apoptosis.[98]

The hypoglycemic action of isolated compound like neohesperidin from *C. aurantium* was studied using rat model of diabetes. It was observed that the experimental group showing decrease in blood glucose and increase in insulin sensitivity in addition to glucose tolerance. These results further confirmed its hypoglycemic actions and possible applications in diabetes treatment.[99]

In conclusion, we can say that *C. aurantium* is well recognized for ages. More study is still required to further discover its novel mechanisms as well as pharmacological actions.

6.2.12.2 CITRUS SINENSIS

The species *C. sinensis* belongs to family Rutaceae also known as sweet orange. In Asia, this fruit has been known for 4000 years. With the transportation and trade expansion routes, orange was familiarized to the world's other regions. It is reported that the sweet orange was present in Liguria, a region in Italy from 1471 to 1472. In 1498, it was introduced into the Iberian Peninsula by Portuguese. In the eighteenth century, an Armenian monk introduced sweet orange to Palestine. It was introduced to Bahia/BR in the centuries of 16th and 18th by Portuguese colonists.

In USA/California, it was introduced by Jesuits where they started their plantations.[100]

The *C. sinensis* species were widely mentioned by Brazilian women community for elementary health. For mild cases of anxiety and insomnia, its leaves were used as a soft tranquilizer in preparation of tea.[101] Furthermore, as a relief for migraine, headache, flu, fever, cough, and allergic reactions, their flowers and fruits were commonly used. In conclusion, we can say that sweet orange is an important medicinal plant usually used by Brazilian community. In conclusion, the sweet orange is one of the most therapeutic plant species utilized by the community of Brazil.[102]

The C. sinensis was traditionally used as a sedative; the active compound responsible for sedation was identified as hesperidin. The sedative action is performed through the interaction and simulation of adenosine receptors. This effect was found opposite to that when coffee or tea is consumed. The state of wakefulness is maintained by the antagonization of the adenosine receptors.[103] In vitro experiments were performed to further investigate the actions of compound hesperidin. It was found that hesperidin possesses antioxidant activity against DPPH free radicals and maximum cytotoxic action against cell lines of human carcinoma. Thus, hesperidin could be a potential future drug. Hesperidin can be extracted from orange peels efficiently with high purity.[104]

When a comparison was done between C. sinensis extracts and positive control through in vitro experiments, the former has shown potent anthelmintic properties. Still this needs to be confirmed through in vivo studies to evaluate the efficacy and for the identification of active components.[105]

Another experiment was performed on rats for 16 weeks to establish the therapeutic application of peel extract for liver fibrosis and cirrhosis. The animals were treated with oral administration of dried peel extract obtained from *C. sinensis* for a time period of 9 weeks in order to evaluate the histopathological as well as biochemical changes related to liver cirrhosis. The curative effect was demonstrated after the examination of the liver tissues. Thus, the orange peel has the potential to be used for curing liver fibrosis and cirrhosis.[106]

The antidiabetic application of the peel extract derived from *C. sinensis* was investigated. In this study, the rats were treated with orange peels at 25 mg/kg concentration with oral administration. The increment in α-amylase action and levels of serum glucose was achieved by administering alloxan at 120 mg/kg concentration (single dose) to the rats. All the negative effects of alloxan were neutralized by giving peels of *C. sinensis. Thus, the antidiabetic properties have been successfully verified.*[107]

In another work done by Faturi et al., the anxiolytic action of the essential oil obtained from *C. sinensis was evaluated by applying* elevated plus maze test on rats. The dwelling time for every rat was noticed in open arms at the time of testing. Further confirmation was done by administering *Melaleuca alternifolia* essential oil but no anxiolytic effect was observed. These rules out the notion that the anxiolytic action was caused by another inhaled fragrance.[108]

In another clinical experiment, a study was conducted to further confirm the anxiolytic property of *C. sinensis. Three groups were selected for study in which two were experimental and one was controlled group. The patients in group one inhaled the essence of essential oil obtained from sweet orange in the waiting area, while another group was allowed to listen to music and the last one was the controlled* group. It was observed that the patient who was intervened with essential oil shows low anxiety level.[109]

Goes et al. in 2012 conducted a study to witness the anxiolytic action of the essential oil obtained from *C. sinensis* on healthy group. A momentous modification in the level of anxiety was observed among the experimental and controlled group. This experiment provides an evident scientific support to confirm the use of *C. sinensis* essential oil for aromatherapy.[110]

In conclusion, we can confidently say that *C. sinensis has a potential against* insomnia and anxiety. The isolated components from *sweet orange have wide spread application in the* pharmaceutical industry as natural antidiabetic, cytotoxic, and anthelmintics agents in addition to useful in liver cirrhosis treatment.

6.2.12.3 CITRUS BERGAMIA

Bergamot is a common name of *Citrus bergamia*[111] which is a member of Rutaceae family belongs to *Esperidea* subfamily and can be defined to be a hybrid between somewhere in *Citrus limon* (lemon) and *Citrus aurantium* (bitter orange).[112] The origin of plant's geographical as well as botanical is not yet clear. Bergamot is thought to be originated from Berga which is a city in Spain from where this plant was exported to Calabria in the state of Southern Italy. *C. bergamia* abundantly practiced in cultivation almost solely grown along with the southern seashore of Calabria. Above 90% of the total global production comes from this very specific region. However, other countries for instance Argentina, Morocco, Iran, Greece, Brazil, and Ivory Coast also cultivate this plant and export the product in a lesser volume.[113]

The *C. bergamia* fruit is primarily employed in the extraction of essential oil, which in turn is hired to make clothing, cosmetics, drugs, perfumes, and food.[114] From cultivation to the process of *C. bergamia*, huge quantity of trash having little marketable value but having high potential of industrial value are produced. Bioactive components, nutrients, and pigments of high level and low toxicity are produced in these wastes.[115]

Evidences suggest that *C. bergamia comprises of* antifungal as well as antibacterial active ingredients along with neuroprotective, neuropsychopharmacological, antiproliferative, and anti-inflammatory in addition to cardiovascular and pain-relieving properties in rodents.[116]

The juice extracted from the endocarp of Bergamot after the essential oil has been obtained, has been reported to have hypolipidemic, anti-inflammatory, and hypoglycemic[117–119] in addition to anticancer effects.[120]

The derivatives found in the essential oil of *C. bergamia* exhibited anti-infective effects against mycetes, larvae, and bacteria in addition to *C. bergamia* juice showed antibacterial effect against *Helicobacter pylori.* Antimicrobial activities have been observed in compounds obtained from the fruit peel extract of *C. bergamia.* The essential oil of *C. bergamia* possess antifungal and antibacterial properties against dermatophytes, *Campylobacter jejuni, Bacillus cereus, Escherichia coli* O157, *Staphylococcus aureus, and Listeria monocytogenes, according to studies.* Its essential oil has *in vitro* actions against the species of *Candida,* indicating that it could be used to treat infections caused by fungal species such as *Candida. In vitro* actions of essential oil obtained from *C. bergamia* against dermatophytes has also been demonstrated.[121]

Additionally, the pharmacological actions of *C. bergamia* in central nervous system syndromes have had positive outcomes. Essential oil of *C. bergamia* derived from the peel is sometimes utilized in the treatment of aromatherapy to combat anxiety as well as stress.[122]

The essential oil of *C. bergamia* was given to rats at varied concentrations and their effects were observed in comparison of diazepam to examine the anxiolytic action of plant. The outcomes of this investigation revealed that the essential oil of *C. bergamia* had an anxiolytic effect, as evidenced by the hole-board and elevated plus maze tests performed on rats. The action of hypothalamic–pituitary–adrenal axis was found to be reduced, resulting in a lower corticosterone response to stress, according to the researchers.[123]

In addition to the effects *C. bergamia* on nervous system, scientific investigations have shown that essential oil obtained from *C. bergamia* has therapeutic effects such as the management of stress response, heart control as well as the reduction of blood pressure.[124]

According to certain reports, the essential oil derived from *C. bergamia* has anticancer properties. Berliocchi et al. found that essential oil of *C. bergamia inhibited the proliferation of* human SH-SY5Y neuroblastoma cells *in vitro*. The toxic action of *C. bergamia* was facilitated in this investigation by numerous pathways activation that result in cell death occurs due to both apoptosis as well as necrosis.[125] The compounds like perillic acid, monoterpenes associated to limonene, limonene, and alcohol obtained from bergamot oil have been shown to suppress human breast cancer cells proliferation in addition to chemotherapeutic as well as chemopreventive properties have observed in mammary tumor models.[126]

Celia et al. established a study regarding the essential oil derived from *C. bergamia* liposomes in 2013, which upgraded the anticancer property against human SH-SY5Y neuroblastoma cells *in vitro* as well as its solubility in water.[127]

In 2014, Di Donna et al. used a model of rat to compare the hypocholesterolemic properties of 3-hydroxy-3-methyl-glutaryl flavanones obtained from the fruit extract of *C. bergamia* compared with simvastatin. LDL (low-density lipoproteins), VLDL (very low-density lipoproteins), HDL (high-density lipoprotein), triglycerides, and total cholesterol level were found to be lowered by administering the derived compound of *C. bergamia* and simvastatin.

Only the derived compound of *C. bergamia*, on the other hand, increased the level of HDL. The derived compound of *C. bergamia* also had not exhibited cytotoxic as well as genotoxic actions at higher concentrations. As a result, the researchers determined that the supplementation of compound derived from *C. bergamia* taken in the diet daily could be hypercholesterolemia properties related to the cardiovascular protecting activities.[128]

Despite the fact that all of *C. bergamia* pharmacological actions point to their future application in medicines, only medical reports' investigation has been reported to yet in the treatment of aromatherapy like anxiolytic action.

6.2.12.4 CITRUS LIMON

C. limon (Linnaeus) N. Burman is the name of lemon originated from Latin. It's a member of the Rutaceae family, and it's also known as limoeiro-azêdo or limoeiro.[129] It is thought to have an origin from Southeast Asia and was brought to Europe by Muslims over the Sicily and Iberian Peninsula. Spain is presently the most important grower of this genus in the Mediterranean area. The third most important species are Lemon which belongs to *Citrus* genus, including wide range natural chemical constituents such as flavonoids, ascorbic acid, citric acid, and minerals.[130]

Some of *C. limon* medicinal activities have recently been documented in the literature. Lemon has been shown in studies to help neutralize the environment for stomach acidity by promoting potassium carbonate synthesis, indicating that it has protective properties on the gastric mucosa. Antiseptic, antipyretic, anti-sclerotic, anti-anemic, analgesic, moisturizer, and emollient activities were also discovered. Lemon cellulose has been shown to have protection of intestinal mucosa, limited hemostatic, antidiarrheal, vascular intoxicant, vascular protector, diuretic as well as vitamin properties.

Lemon synthesis networks produce a wide range of by-products and wastes, which are a good source of natural biological active constituents that could be used in processed foods, animal feed as well as healthcare. Though its health advantages have long been linked to the content of its vitamin C. Flavonoids are recently discovered having an essential role. Some researchers believe that the occurrence of flavonoids in lemon possess antiviral, antiproliferative, anticancer, antimutagenic, antioxidant, antiallergic, and anti-inflammatory properties.

The major flavonoid found in *C. limon* is Hesperidin which raises the capillary resistance, affects the vascular permeability, anti-inflammatory, and analgesic activities. It's also a good antioxidant because it is able to scavenge cancer causing free radicals. Flavonoids isolated from lemon juice have hypocholesterolemic effects, according to certain research.[131]

In year 1996, an experiment conducted by Trovato et al. on rats and discovered that *C. limon* had remarkable action on triglycerides and cholesterol levels, signifying that the use of its juice for a long term could provide better hypercholesterolemia protection.[132]

The essential oil obtained from *C. limon* was used in other report to protect Wistar albino rats against renal damage and acute hepatic caused by administering high dose of aspirin. The findings of this investigation revealed that *C. limon therapy* protected the kidney as well as liver against aspirin-induced damage.[133]

In terms of pharmacological properties on the central nervous system, Khan and Riaz recently investigated to check the lemon effects on the rat's behavior by utilizing three different doses of 0.2, 0.4, and 0.6 mL/kg, respectively, which were classified as low, moderate, and high doses. Elevated plus maze, open field and forced swimming tests were used to assess the antidepressant and anxiolytic actions were two times over a 15 days period. At a moderate dose, *C. limon showed* an increase in distance travelled, rearing number, and central entries in the open field test, whereas open arm entries number was observed to be enhanced in the elevated plus maze test. While there was found an increase in climbing period and decrease in immobility duration in the forced swimming test. As a results, the findings imply that *C. limon* has an anxiolytic property at moderate dose.[134]

It's been noticed that the diseases like depression and anxiety can be achieved by improved dietary changes, since evidence suggests that a food high in vitamins and antioxidants decreases these signs. An experiment was conducted to investigate the potential actions of *Punica granatum and C. limon on behavior of rats.* In this investigation, two combined doses of *P. granatum and C. limon* were utilized: 0.2 + 8 and 0.4 + 5 mL/kg, respectively. The anxiolytic and antidepressant action were assessed two times, for a total of 15 days with the help of forced swimming, open field as well as elevated plus maze tests, as in the prior study. At the combined doses of 0.4 + 5 mL/kg, the utilization of *P. granatum and C. limon* increased the number of entries and distance traveled in open field test. At higher combination dose of 0.2 + 8 mL/kg, the number of central entries was raised in elevated plus maze test. At both combination doses of 0.2 + 8 mL/kg as well as 0.4 + 5 mL/kg, the duration of climbing was increased while the duration of immobility was decreased in forced swimming test. The researchers concluded on behalf of these findings that a combined dose of 0.4 + 5 mL/kg of *P. granatum as well as C. limon* showed anti-depressant and anxiolytic properties.[135]

Riaz et al. conducted another experiment in 2014 to check how juice of pomegranate and *C. limon* affected the memory of rats. The quality of sleep, diet, and stress are all known to have a significant impact on memory. *C. limon* comprises of important nutrients as well as phytochemicals that promote memory, the memory of short-term in particular, according to the findings of this study. They also came to the conclusion that these effects are due to the occurrence of flavonoids in their juices.[136]

A compound Hesperidin isolated from *C. limon, was discovered to have a significant* vasodilating effect. Dobias et al. studied the activities of hesperidin compound on responses of vascular system in normal rats as well as hypertensive in 2016. There is no effect of Hesperidin was observed on blood pressure in SHR and Wistar rats, but it greatly increased vasodilator response of endothelium-dependent. All of the groups increased the responses of contraction at same level but only SHR group was significantly reduced the responses of relaxation. Only in SHR rats given hesperidin did significantly blocked the channels of potassium impair the vasodilation of endothelium-dependent. This suggests that hesperidin can help hypertensive patients with vasodilation of endothelium-dependent.[137]

In 2016, Amorim et al. published a study of antinociceptive and anti-inflammatory properties showed by the EO (essential oil) extracted from *C. latifolia*, *C. limonia, C. aurantifolia*, and *C. limon.*[138]

Vaiyapuri et al. reported an experiment in 2015 to validate the pharmacological properties of polymethoxy flavonoid *C. limon such as* nobiletin which helps in the regulation of platelet. Nobiletin derived from dietary source also reported for potential antithrombotic effect.[139]

Finally, *C. limon* has a wide range of medicinal characteristics, and it has long been traditionally used as a medicine to cure a variety of ailments. More research is needed, nevertheless, before it can be established and spread its wide applications in medical preparations.[140]

6.3 SUMMARY

A wide range of drug lead compounds can be found in medicinal plants. The plants or their various parts belong to the family Rutaceae, which is

TABLE 6.1 Pharmacological Properties of Various Species of Family Rutaceae (*In Vitro* and *In Vivo Studies*).

S. No	Species	Family	Pharmacological action
1.	*Fagara leprieurii*	Rutaceae	Cancer, kidney ache, ulcer, gonorrhea, gingivitis, gastritis, bilharzia, sterility, diarrhea, laxative and other various pathogenic diseases.
2.	*Fraxinus xanthoxyloides*	Rutaceae	Antiseptic, laxative and have antifungal properties.
3.	*Micromelum minutum*	Rutaceae	Antileishmanicidal, anticancer, regulation of menstruation, malaise, and tumor.
			Leaves (general tonic): hemorrhoids, sore tongue, cough, headache, white scum on tongue, toothache, teething problems in babies, bad breath, skin allergies, scabies, gonorrhoea, stomachache, and thrush.
			Shoots (used as medicine): infantile convulsions. Stems: as a carminative.
			Flowers and fruits: as an expectorant and a purgative.
			Roots: as a febrifuge, carminative, diarrhea in children.
4.	*Zanthoxyli fructus*	Rutaceae	Relieves pain (abdominal pain), expels cold sensation in the abdomen and known for antiparasitic, mildly diuretic, analgesic, antibiotic, and antidiarrhea.
5.	*Ruta graveolens*	Rutaceae	Human brain cancer cells, pain, inflammatory diseases, dermatitis, DNA strand breakage protection, mutagenesis, and rheumatism.
6.	*Phellodendron amurense*	Rutaceae	Antibiotic, antihypertensive, antitussive, expectorant, antiasthmatic, dysentery, chronic bronchitis, meningitis, and conjunctivitis in children.
			Topical application: trichomonas cervicitis, vaginitis, and eczema on ear.
7.	*Murraya paniculate*	Rutaceae	Pain, stimulant, astringent effect, dysentery, and diarrhea.
			Roots: for coughs, rheumatism, and hysteria.
			Cooked leaves and boiled twigs: for stomachache and inflamed joints.
			Known for antinociceptive, antioxidant, antidiabetic, antimicrobial, and analgesic activities.

TABLE 6.1 *(Continued)*

S. No	Species	Family	Pharmacological action
8.	*Aegle marmelos*	Rutaceae	General medicine: for smallpox, tumors, epilepsy, leprosy, eye disorders, thyroid disorders, jaundice, myalgia, dysentery, constipation, ulcers, stomachic, indigestion, thirst, stomachache, acidity, chronic diarrhea, abdominal discomfort, nausea, fever, snakebite, sores, asthma, upper respiratory tract infections, acute bronchitis, swelling, inflammations, burning sensation, febrile delirium, spermatorrhea, mental illnesses, and leukoderma.
			Known for antioxidant, antidiarrheal, antimicrobial, and antifertility activities.
9.	*Zanthoxylum armatum*	Rutaceae	Used as carminative, stomachic, and teeth cleaning.
			Aromatic tonic: in dyspepsia and fever.
			Aromatic lotion: for scabies, antiseptic, and disinfectant action.
			Known for antibacterial, antifungal, antiviral, hypoglycaemic, anthelmintic, insecticidal, antioxidant, hepatotoxicity, anti-inflammatory, and inhibition of tumor-necroting factor
10.	*Glycosmis pentaphylla*	Rutaceae	Antibacterial, anthelmintic, antifungal, Anti-oxidant, anti-arsenicosis, mosquitocidal, antidiabetic, analgesic, antipyretic, anti-inflammatory, wound healing, antihyperlipidemic, anticancer, and cytotoxic properties.
11.	*Citrus aurantium* L.	Rutaceae	General tonic, gastrointestinal stimulant, hypoglycemic effect, anxiolytic effect, nervous system illnesses, anxiety, sleeplessness, hysteria, acidity of stomach, stomach cramps, constipation, and anti-inflammatory.
12.	*Citrus sinensis*	Rutaceae	Anxiolytic, liver cirrhosis, anthelmintic, sedative, and antidiabetic effects.
13.	*Citrus bergamia*	Rutaceae	Anxiolytic activity, antiproliferative, neuroprotective, analgesic, anti-inflammatory, anticancer, antifungal, antibacterial, hypolipidemic, and hypoglycemic properties.
14.	*Citrus limon*	Rutaceae	Antiproliferative, anticancer, antimutagenic, analgesic, moisturizer, Emollient, antiallergic, anti-anemic, antipyretic, antisclerotic, antiseptic, diuretic, Intestinal mucosa protector, antidiarrheal, anti-inflammatory, antiviral, antioxidant, local hemostatic and vascular stimulant and protector.

known for their valuable medicinal and therapeutic properties. Secondary metabolites such as flavonoids, alkaloids, coumarins, limonoids, volatile oils, and essential oils isolated from different parts of plants can be found in abundance in the Rutaceae family and all of them having physiological and pharmacological properties. This current review provides data regarding the compounds isolated from various species of Rutaceae family and their physiological and pharmacological activities. The diversity of chemical constituents found in various species of Rutaceae family demonstrates a wide range of therapeutic and pharmacological properties. With existing information revealing potential health applications and therapeutic uses of family Rutaceae, this review promotes further investigation of other species of the family Rutaceae whose chemical and biological assays have yet to be studied.

KEYWORDS

- **medicinal plants**
- **Rutaceae family**
- **secondary metabolites**
- **physiological**
- **pharmacological**

REFERENCES

1. Fatema-Tuz-Zohora, H. C.; Ahsan, M. Chemical Constituents, Cytotoxic Activities and Traditional Uses of *Micromelum minutum* (Rutaceae): A Review. *Pharm. Pharmacol. Int. J.* **2019,** *7* (5), 229–236.
2. Supabphol, R.; Tangjitjareonkun, J. Chemical Constituents and Biological Activities of *Zanthoxylum limonella* (Rutaceae): A Review. *Trop. J.Pharma. Res.* **2014,** *13* (12), 2119–2130.
3. Lewis, J. R. Biological Activity of Some Rutaceous Compounds. *Chemistry and Chemical Taxonomy of Rutales*; Academic Press: London, 1983; pp 301–318.
4. Ali, M. S.; Fatima, S.; Pervez, M. K. Haplotin: A New Furanoquinoline from *Haplophyllum acutifolium* (Rutaceae). *J. Chem. Soc. Pak.* **2008,** *30*, 775–779.
5. Mandal, S.; Nayak, A.; Kar, M.; Banerjee, S. K.; Das, A.; Upadhyay, S. N.; Banerji, J. Antidiarrhoeal Activity of Carbazole Alkaloids from Murraya Koenigii Spreng (Rutaceae) Seeds. *Fitoterapia* **2010,** *81* (1), 72–74.

6. Cardoso-Lopes, E. M.; Maier, J. A.; Silva, M. R. D.; Regasini, L. O.; Simote, S. Y.; Lopes, N. P. et al. Alkaloids from Stems of *Esenbeckia leiocarpa* Engl. (Rutaceae) as Potential Treatment for Alzheimer Disease. *Molecules* **2010,** *15* (12), 9205–9213.
7. Barrera, C. A. C.; Barrera, E. D. C.; Falla, D. S. G.; Murcia, G. D.; Suarez, L. E. C. Seco-Limonoids and Quinoline Alkaloids from *Raputia heptaphylla* and Their Antileishmanial Activity. *Chem. Pharma. Bull.* **2011,** *59* (7), 855–859.
8. Rajkumar, S.; Jebanesan, A. Bioactivity of Flavonoid Compounds from *Poncirus trifoliata* L. (Family: Rutaceae) Against the Dengue Vector, *Aedes aegypti* L. (Diptera: Culicidae). *Parasitol. Res.* **2008,** *104* (1), 19–25.
9. Emam, A. M.; Swelam, E. S.; Megally, N. Y. Furocoumarin and Quinolone Alkaloid with Larvicidal and Antifeedant Activities Isolated from Ruta chalepensis Leaves. *J. Nat. Prod.* **2009,** *2*, 10–22.
10. Su, J. F.; Guo, C. J.; Wei, J. Y.; Yang, J. J.; Jiang, Y. G.; Li, A. Y. F. Protection Against Hepatic Ischemia-Reperfusion Injury in Rats by Oral Pretreatment with Quercetin. *Biomed. Environ. Sci.* **2003,** *16* (1), 1–8.
11. Swingle, W. T.; Reece, P. C. The Botany of Citrus and Relatives. In *The Citrus Industry*; University of California: Berkeley, 1967.
12. Huang, S.; Wang, J. H.; Luo, X. M. Research Process on Chemical Constituents and Pharmacological Activities of Micromelum [J]. *Chin. Med. Mat.* **2011,** *34* (10), 1635–1638.
13. Zhi-Yao, W. A. N. G.; Wen-Jun, H. E.; Wen-Bing, Z. H. O. U.; Guang-Zhi, Z. E. N. G.; Zhi-Qi, Y. I. N.; Shou-Xun, Z. H. A. O.; Ning-Hua, T. A. N. Two New Phenylpropanoids from *Micromelum integerrimum. Chin. J. Nat. Med.* **2014,** *12* (8), 619–622.
14. Koriem, K. M.; Aminuddin, M. E.; Kader, A. S.; Sheikh, N. R. Antihyperglycemic, Antihyperlipidemic and Antiapoptotic Activities of Micromelumminutum Seeds in Diabetic Rats. *J. Mol. Genet. Med.* **2013,** 2–8.
15. Hyland, B. P. M.; Whiffin, T.; Zich, F. A. *Australian Tropical Rainforest Plants*, ed. 6.1; Online Version [RFK 6.1], 2010.
16. Wiart, C. *Medicinal Plants of Asia and the*; CRC Press, 2006; pp 223–224 p.
17. Cambie, R. C.; Ash, J. *Fijian Medicinal Plants*; CSIRO Publishing, 1994; pp 2031–2032 p.
18. Kuete, V.; Krusche, B.; Youns, M.; Voukeng, I.; Fankam, A. G.; Tankeo, S.; Efferth, T. Cytotoxicity of Some Cameroonian Spices and Selected Medicinal Plant Extracts. *J. Ethnopharmacol.* **2011,** *134* (3), 803–812.
19. Ngane, A. N.; Biyiti, L.; Zollo, P. A.; Bouchet, P. H. Evaluation of Antifungal Activity of Extracts of Two Cameroonian Rutaceae: Zanthoxylum leprieurii Guill. et Perr. and Zanthoxylum xanthoxyloides Waterm. *J. Ethnopharmacol.* **2000,** *70* (3), 335–342.
20. Croft, K. D.; Toia, R. F. Coumarins from *Micromelum minutum. Planta Medica* **1989,** *55* (04), 401–401.
21. Itharat, A.; Houghton, P. J.; Eno-Amooquaye, E.; Burke, P. J.; Sampson, J. H.; Raman, A. In Vitro Cytotoxic Activity of Thai Medicinal Plants Used Traditionally to Treat Cancer. *J. Ethnopharmacol.* **2004,** *90* (1), 33–38.
22. Fatema-Tuz-Zohora, H. C.; Ahsan, M. Chemical Constituents, Cytotoxic Activities and Traditional Uses of *Micromelum minutum* (Rutaceae): A Review. *Pharm. Pharmacol. Int. J.* **2019,** *7* (5), 229–236.

23. Fern, K. *Plants for a Future: Edible & Useful Plants for a Healthier World*; Permanent Publications, 1997.
24. Roy, M. K.; Thalang, V. N.; Trakoontivakorn, G.; Nakahara, K. Mahanine, a Carbazole Alkaloid from Micromelum Minutum, Inhibits Cell Growth and Induces Apoptosis in U937 Cells Through a Mitochondrial Dependent Pathway. *Br. J. Pharmacol.* **2005,** *145* (2), 145–155.
25. Croft, K. D.; Toia, R. F. Coumarins from *Micromelum minutum. Planta medica* **1989,** *55* (04), 401–401.
26. Tantishaiyakul, V.; Pummangura, S.; Chaichantipyuth, C.; Ma, W. W.; McLaughlin, J. L. Phebalosin from the Bark of *Micromelum minutum. J. Nat. Products* **1986,** *49* (1), 180–181.
27. Lekphrom, R.; Kanokmedhakul, S.; Kukongviriyapan, V.; Kanokmedhakul, K. C-7 Oxygenated Coumarins from the Fruits of *Micromelum minutum. Arch. Pharmacal. Res.* **2011,** *34* (4), 527–531.
28. Rahmani, M.; Susidarti, R. A.; Ismail, H. B.; Sukari, M. A.; Hin, T. Y. Y.; Lian, G. E. C.; Waterman, P. G. Coumarins from Malaysian *Micromelum minutum. Phytochemistry* **2003,** *64* (4), 873–877.
29. Rahmani, M.; Taufiq-Yap, Y. H.; Ismail, H. B.; Sukari, A.; Waterman, P. G. New Coumarin and Dihydrocinnamic Acid Derivatives from Two Malaysian Populations of *Micromelum minutum. Phytochemistry* **1994,** *37* (2), 561–564.
30. Asmah Susidarti, R.; Rahmani, M.; Ismail, H. B.; Sukari, M. A.; Yun Hin, T. Y.; Ee Cheng Lian, G.; Waterman, P. G. A New Coumarin and Triterpenes from Malaysian *Micromelum minutum. Nat. Product Res.* **2006,** *20* (2), 145–151.
31. Skehan, P.; Storeng, R.; Scudiero, D.; Monks, A.; McMahon, J.; Vistica, D.; et al. New Colorimetric Cytotoxicity Assay for Anticancer-Drug Screening. *J. Natl. Cancer Inst.* **1990,** *82* (13), 1107–1112.
32. Sharker, S. M.; Shahid, I. J.; Hasanuzzaman, M. Antinociceptive and Bioactivity of Leaves of *Murraya paniculata* (L.) Jack, Rutaceae. *Revista Brasileira de Farmacognosia* **2009,** *19* (3), 746–748.
33. Mondal, S. K.; Ray, B.; Ghosal, P. K.; Teleman, A.; Vuorinen, T. Structural Features of a Water Soluble Gum Polysaccharide from *Murraya paniculata* Fruits. *Int. J. Biol. Macromol.* **2001,** *29*, 169–174.
34. Chatterjee, A.; Pakrashi, S. C. *The Treatise on Indian Medicinal Plants* (reprinted ed.); Publications and Information Directorate: New Delhi, 2001.
35. Gautam, M. K.; Gupta, A.; Vijaykumar, M.; Rao, C. V.; Goel, R. K. Studies on the Hypoglycemic Effects of *Murraya paniculata* Linn. Extract on Alloxan-Induced Oxidative Stress in Diabetic and Non-Diabetic Models. *Asian Pac. J. Trop. Dis.* **2012,** *2*, S186–S191.
36. Sayar, K.; Paydar, M.; Pingguan-Murphy, B. Pharmacological Properties and Chemical Constituents of *Murraya paniculata* (L.) Jack. *Med. Aromatic Plants* **2014,** *3* (4), 1–6.
37. Rohman, A.; Riyanto, S. Antioxidant Potency of Ethanolic Extract of Kemuning Leaves (*Murraya paniculata* (L) Jack) In Vitro. *Indonesian J. Pharm.* **2005,** 136–140.
38. Zhang, J. Y.; Li, N.; Che, Y. Y.; Zhang, Y.; Liang, S. X.; Zhao, M. B. et al. Characterization of Seventy Polymethoxylated Flavonoids (PMFs) in the Leaves of *Murraya paniculata* by On-Line High-Performance Liquid Chromatography Coupled

to Photodiode Array Detection and Electrospray Tandem Mass Spectrometry. *J. Pharma. Biomed. Anal.* **2011,** *56* (5), 950–961.

39. Podder, M. K.; Das, B. N.; Saha, A.; Ahmed, M. Analgesic Activity of Bark of *Murraya paniculata. Int. J. Med. Med. Sci.* **2011,** *3* (4), 105–108.
40. Sayar, K.; Paydar, M.; Pingguan-Murphy, B. Pharmacological Properties and Chemical Constituents of *Murraya paniculata* (L.) Jack. *Med. Aromatic Plants* **2014,** *3* (4), 1–6.
41. Barros, F. A. P.; Rodrigues-Filho, E. Four Spiroquinazoline Alkaloids from Eupenicillium sp. Isolated as an Endophytic Fungus from Leaves of *Murraya paniculata* (Rutaceae). *Biochem. System. Ecol.* **2005,** *33* (3), 257–268.
42. Li, Q.; Zhu, L. F.; But, P. P. H.; Kong, Y. C.; Chang, H. T.; Waterman, P. G. Monoterpene and Sesquiterpene Rich Oils from the Leaves of Murraya Species: Chemotaxonomic Significance. *Biochem. System. Ecol.* **1988,** *16* (5), 491–494.
43. Ito, C.; Furukawa, H.; Ishii, H.; Ishikawa, T.; Haginiwa, J. The Chemical Composition of *Murraya paniculata*. The Structure of Five New Coumarins and One New Alkaloid and the Stereochemistry of Murrangatin and Related Coumarins. *J. Chem. Soc., Perkin Trans.* **1990,** *1*, (7), 2047–2055.
44. Kinoshita, T.; Firman, K. Myricetin 5, 7, 3′, 4′, 5′-pentamethyl Ether and Other Methylated Flavonoids from *Murraya paniculata. Phytochemistry* **1997,** *45* (1), 179–181.
45. Ferracin, R. J.; da Silva, M. F. D. G.; Fernandes, J. B.; Vieira, P. C. Flavonoids from the Fruits of *Murraya paniculata. Phytochemistry* **1998,** *47* (3), 393–396.
46. Sekar, D. K.; Kumar, G.; Karthik, L.; Rao, K. B. A Review on Pharmacological and Phytochemical Properties of *Aegle marmelos* (L.) Corr. Serr. (Rutaceae). *Asian J. Plant Sci. Res.* **2011,** *1* (2), 8–17.
47. Sivaraj, R.; Balakrishnan, A.; Thenmozhi, M.; Venckatesh, R. Antimicrobial Activity of *Aegle marmelos*, *Ruta graveolens*, *Opuntia dellini*, *Euphorbia royleena* and *Euphorbia antiquorum*. *J. Pharm. Res.* **2011,** *4* (5), 1507.
48. Joshi, P. V.; Patil, R. H.; Maheshwari, V. L. In Vitro Antidiarrhoeal Activity and Toxicity Profile of *Aegle marmelos* Correa ex Roxb. Dried Fruit Pulp, 2009.
49. Sathiyaraj, K.; Sivaraj, A.; Madhumitha, G.; Kumar, P. V.; Saral, A. M.; Devi, K.; Kumar, B. S. Antifertility Effect of Aqueous Leaf Extract of *Aegle marmelos* on Male Albino Rats. *Int. J. Curr. Pharma. Res.* **2010,** *2*, 26–29.
50. Paul, A.; Kumar, A.; Singh, G.; Choudhary, A. Medicinal, Pharmaceutical and Pharmacological Properties of *Zanthoxylum armatum*: A Review. *J. Pharmacogn. Phytochem.* **2018,** *7* (4), 892–900.
51. Islam, M. S.; Akhtar, M. M.; Rahman, M. M.; Rahman, M. A.; Sarker, K. K.; Alam, M. F. Antitumor and Phytotoxic Activities of Leaf Methanol Extract of *Oldenlandia diffusa* (Willd.) Roxb. *Global J. Pharmacol.* **2009,** **3** (2), 99–106.
52. Tiwary, M.; Naik, S. N.; Tewary, D. K.; Mittal, P. K.; Yadav, S. Chemical Composition and Larvicidal Activities of the Essential Oil of *Zanthoxylum armatum* DC (Rutaceae) Against Three Mosquito Vectors. *J. Vector Borne Dis.*, *44* (3), 198.
53. Sati, S. C.; Sati, M. D.; Raturi, R.; Badoni, P.; Singh, H. Anti-Inflammatory and Antioxidant Activities of *Zanthoxylum armatum* Stem Bark. *Glob. J. Res. Eng.* **2011,** *11* (5), 19–22.

54. Verma, N.; Khosa, R. L. Hepatoprotective Activity of Leaves of *Zanthoxylum armatum* DC in CCl 4 Induced Hepatotoxicity in Rats, 2010.
55. Peana, A. T.; D'Aquila, P. S.; Panin, F.; Serra, G.; Pippia, P.; Moretti, M. D. L. Anti-Inflammatory Activity of Linalool and Linalyl Acetate Constituents of Essential Oils. *Phytomedicine* **2002,** *9* (8), 721–726.
56. Ghani, A. *Medicinal Plants of Bangladesh: Chemical Constituents and Uses*; Asiatic Society of Bangladesh, 1998.
57. Bowen, I. H.; Perera, K. C.; Lewis, J. R. Alkaloids of the Leaves of *Glycosmis bilocularis*. *Phytochemistry* **1978,** *17* (12), 2125–2127.
58. Ganesan, S. Traditional Oral Care Medicinal Plants Survey of Tamil Nadu. *Nat Prod. Radiance* **2008,** *7*, 166–172.
59. Chakravarti, D.; Chakravarti, R. N.; Cohen, L. A.; Dasgupta, B.; Datta, S.; Miller, H. K. Alkaloids of *Glycosmis arborea*—II: Structure of Arborine. *Tetrahedron* **1961,** *16* (1–4), 224–250.
60. Chakravarty, A. K.; Sarkar, T.; Masuda, K.; Shiojima, K. Carbazole Alkaloids from Roots of *Glycosmis arborea*. *Phytochemistry* **1999,** *50* (7), 1263–1266.
61. Jeyachandran, R.; Mahesh, A.; Cindrella, L.; Sudhakar, S.; Pazhanichamy, K. Antibacterial Activity of Plumbagin and Root Extracts of *Plumbago zeylanica* L. *Acta Biol. Cracoviensia Series Bot.* **2009,** *51* (1), 17–22.
62. Lewis, S.; Tarrier, N.; Haddock, G.; Bentall, R.; Kinderman, P.; Kingdon, D.; Dunn, G. Randomised Controlled Trial of Cognitive-Behavioural Therapy in Early Schizophrenia: Acute-Phase Outcomes. *Br. J. Psych.* **2002,** *181* (S43), s91-s97.
63. Lattaoui, N. Individual and Combined Antibacterial Activity of the Main Constituents of Three Essential Oils. *Eur. Mediterranean Plant Protect. Organ. Bull.* **1994,** *3*, 13–19.
64. Campos, F. M.; Couto, J. A.; Figueiredo, A. R.; Tóth, I. V.; Rangel, A. O.; Hogg, T. A. Cell Membrane Damage Induced by Phenolic Acids on Wine Lactic Acid Bacteria. *Int. J. Food Microbiol.* **2009,** *135* (2), 144–151.
65. Bajpai, V. K.; Al-Reza, S. M.; Choi, U. K.; Lee, J. H.; Kang, S. C. Chemical Composition, Antibacterial and Antioxidant Activities of Leaf Essential Oil and Extracts of *Metasequioa glyptostroboides* Miki ex Hu. *Food Chem. Toxicol.* **2009,** *47* (8), 1876–1883.
66. Shin, S. Y.; Bajpai, V. K.; Kim, H. R.; Kang, S. C. Antibacterial Activity of Bioconverted Eicosapentaenoic (EPA) and Docosahexaenoic Acid (DHA) Against Foodborne Pathogenic Bacteria. *Int. J. Food Microbiol.* **2007,** *113* (2), 233–236.
67. Paul, S.; Dubey, R. C.; Maheswari, D. K.; Kang, S. C. *Trachyspermum ammi* (L.) Fruit Essential Oil Influencing on Membrane Permeability and Surface Characteristics in Inhibiting Food-Borne Pathogens. *Food Control* **2011,** *22* (5), 725–731.
68. Sharma, A.; Bajpai, V. K.; Baek, K. H. Determination of Antibacterial Mode of Action of A llium sativum Essential Oil against Foodborne Pathogens Using Membrane Permeability and Surface Characteristic Parameters. *J. Food Saf.* **2013,** *33* (2), 197–208.
69. Zou, Z.; Xi, W.; Hu, Y.; Nie, C.; Zhou, Z. Antioxidant Activity of Citrus Fruits. *Food Chem.* **2016,** *196*, 885–896.
70. González, C. A.; Sala, N.; Rokkas, T. Gastric Cancer: Epidemiologic Aspects. *Helicobacter* **2013,** *18*, 34–38.

71. Pimenta, F. C. F.; Alves, M. F.; Pimenta, M. B. F.; Melo, S. A. L.; Almeida, A. A. F. D.; Leite, J. R. et al. Anxiolytic Effect of *Citrus aurantium* L. on Patients with Chronic Myeloid Leukemia. *Phytother. Res.* **2016,** *30* (4), 613–617.
72. Md Othman, S. N. A.; Hassan, M. A.; Nahar, L.; Basar, N.; Jamil, S.; Sarker, S. D. Essential Oils from the Malaysian Citrus (Rutaceae) Medicinal Plants. *Medicines* **2016,** *3* (2), 13.
73. Palazzolo, E.; Laudicina, V. A.; Germanà, M. A. Current and Potential Use of Citrus Essential Oils. *Curr. Org. Chem.* **2013,** *17* (24), 3042–3049.
74. Wolffenbüttel, A. N.; Zamboni, A.; Becker, G.; Dos Santos, M. K.; Borille, B. T.; de Cássia Mariotti, K. et al. Citrus Essential Oils Inhalation by Mice: Behavioral Testing, GCMS Plasma Analysis, Corticosterone, and Melatonin Levels Evaluation. *Phytother. Res.* **2018,** *32* (1), 160–169.
75. Lawless, J. The Complete Illustrated Guide to Aromatherapy: A Practical Approach to the Use of Essential Oils for Health and Well-Being. *Element* **1999**.
76. Conforti, F.; Statti, G. A.; Tundis, R.; Loizzo, M. R.; Menichini, F. In vitro Activities of *Citrus medica* L. cv. Diamante (Diamante Citron) Relevant to Treatment of Diabetes and Alzheimer's Disease. *Phytother. Res.* **2007,** *21* (5), 427–433.
77. Cheng, S. S.; Chung, M. J.; Lin, C. Y.; Wang, Y. N.; Chang, S. T. Phytochemicals from *Cunninghamia konishii* Hayata Act as Antifungal Agents. *J. Agric. Food Chem.* **2012,** *60* (1), 124–128.
78. Sheikh, M.; Safiuddin, A.; Khan, Z.; Rizvi, R.; Mahmood, I. Antibacterial and Antifungal Potential of Some Medicinal Plants Against Certain Phytopathogenic Micro-Organisms. *Arch. Phytopathol. Plant Protect.* **2013,** *46* (9), 1070–1080.
79. Velázquez-Nuñez, M. J.; Avila-Sosa, R.; Palou, E.; López-Malo, A. Antifungal Activity of Orange (*Citrus sinensis* var. Valencia) Peel Essential Oil Applied by Direct Addition or Vapor Contact. *Food Control* **2013,** *31* (1), 1–4.
80. Souza, E. L. D.; Lima, E. D. O.; Freire, K. R. D. L.; Sousa, C. P. D. Inhibitory Action of Some Essential Oils and Phytochemicals on the Growth of Various Moulds Isolated from Foods. *Braz. Arch. Biol. Technol.* **2005,** *48*, 245–250.
81. Chanthaphon, S.; Chanthachum, S.; Hongpattarakere, T. Antimicrobial Activities of Essential Oils and Crude Extracts from Tropical Citrus spp. Against Food-Related Microorganisms. *Songklanakarin J. Sci. Technol.* **2008,** *30*.
82. Rammanee, K.; Hongpattarakere, T. Effects of Tropical Citrus Essential Oils on Growth, Aflatoxin Production, and Ultrastructure Alterations of *Aspergillus flavus* and *Aspergillus parasiticus*. *Food Bioprocess Technol.* **2011,** *4* (6), 1050–1059.
83. Fisher, K.; Phillips, C. Potential Antimicrobial Uses of Essential Oils in Food: Is Citrus the Answer? *Trends Food Sci. Technol.* **2008,** *19* (3), 156–164.
84. Calo, J. R.; Crandall, P. G.; O'Bryan, C. A.; Ricke, S. C. Essential Oils as Antimicrobials in Food Systems–A Review. *Food Control* **2015,** *54*, 111–119.
85. Ramos, C. A. F.; Sá, R. D. C. D. S.; Alves, M. F.; Benedito, R. B.; de Sousa, D. P.; Margareth de Fátima, F. M. et al. Histopathological and Biochemical Assessment of D-Limonene-Induced Liver Injury in Rats. *Toxicol. Rep.* **2015,** *2*, 482–488.
86. Arias, B. A.; Ramón-Laca, L. Pharmacological Properties of Citrus and Their Ancient and Medieval Uses in the Mediterranean Region. *J. Ethnopharmacol.* **2005,** *97* (1), 89–95.

87. Bouchard, N. C.; Howland, M. A.; Greller, H. A.; Hoffman, R. S.; Nelson, L. S. Ischemic Stroke Associated with Use of an Ephedra-Free Dietary Supplement Containing Synephrine. *Mayo Clin. Proc.* **2005,** *80* (4), 541–545.
88. De Moraes Pultrini, A.; Galindo, L. A.; Costa, M. Effects of the Essential Oil from *Citrus aurantium* L. in Experimental Anxiety Models in Mice. *Life Sci.* **2006,** *78* (15), 1720–1725.
89. Silva, M. D.; Dreveck, S.; Zeni, A. L. B. Ethnobotanical Study of Medicinal Plants Used by Rural Population Around the Itajaí National Park. *Indaial* **2009,** *10* (2), 54–64.
90. Pimenta, F. C. F.; Cunha-Tavares, N. A.; Chaves-Neto, G.; Alves, M.; Fernandes-Pimenta, M.; Melo-Diniz, J. et al. Pharmacological Actions of Citrus Species. *Citrus Pathol.*; Gill, H., Garg, H., Eds., 2017; pp 197–211.
91. Carvalho-Freitas, M. I. R.; Costa, M. Anxiolytic and Sedative Effects of Extracts and Essential Oil from *Citrus aurantium* L. *Biol. Pharma. Bull.* **2002,** *25* (12), 1629–1633.
92. Leite, M. P.; Fassin Jr, J.; Baziloni, E. M.; Almeida, R. N.; Mattei, R.; Leite, J. R. Behavioral Effects of Essential Oil of *Citrus aurantium* L. Inhalation in Rats. *Revista Brasileira de Farmacognosia* **2008,** *18*, 661–666.
93. Costa, C. A.; Cury, T. C.; Cassettari, B. O.; Takahira, R. K.; Flório, J. C.; Costa, M. Citrus Aurantium L. Essential Oil Exhibits Anxiolytic-Like Activity Mediated by 5-HT 1A-Receptors and Reduces Cholesterol After Repeated Oral Treatment. *BMC Complement. Altern. Med.* **2013,** *13* (1), 1–10.
94. Akhlaghi, M.; Shabanian, G.; Rafieian-Kopaei, M.; Parvin, N.; Saadat, M.; Akhlaghi, M. *Citrus aurantium* Blossom and Preoperative Anxiety. *Revista brasileira de Anestesiologia* **2011,** *61*, 707–712.
95. Namazi, M.; Amir AliAkbari, S.; Mojab, F.; Talebi, A.; Alavi Majd, H.; Jannesari, S. Investigating the Effects of Aromatherapy with *Citrus aurantium* Oil on Anxiety During the First Stage of Labor. *Iran. J. Obstetr. Gynecol. Infert.* **2014,** *17* (111), 12–19.
96. Namazi, M.; Akbari, S. A. A.; Mojab, F.; Talebi, A.; Majd, H. A.; Jannesari, S. Effects of *Citrus aurantium* (Bitter Orange) on the Severity of First-Stage Labor Pain. *Iran. J. Pharma. Res.* **2014,** *13* (3), 1011.
97. Kang, S. R.; Park, K. I.; Park, H. S.; Lee, D. H.; Kim, J. A.; Nagappan, A. et al. Anti-Inflammatory Effect of Flavonoids Isolated from Korea *Citrus aurantium* L. on Lipopolysaccharide-Induced Mouse Macrophage RAW 264.7 Cells by Blocking of Nuclear Factor-Kappa B (NF-κB) and Mitogen-Activated Protein Kinase (MAPK) Signalling Pathways. *Food Chem.* **2011,** *129* (4), 1721–1728.
98. Park, K. I.; Park, H. S.; Kim, M. K.; Hong, G. E.; Nagappan, A.; Lee, H. J. et al. Flavonoids Identified from Korean *Citrus aurantium* L. Inhibit Non-Small Cell Lung Cancer Growth In Vivo and In Vitro. *J. Funct. Foods* **2014,** *7*, 287–297.
99. Jia, S.; Hu, Y.; Zhang, W.; Zhao, X.; Chen, Y.; Sun, C. et al. Hypoglycemic and Hypolipidemic Effects of Neohesperidin Derived from *Citrus aurantium* L. in Diabetic KK-A y Mice. *Food Funct.* **2015,** *6* (3), 878–886.
100. Donadio LC. Orange pear. Jaboticabal: FUNEP; 1999.
101. Nazária MDE, Campelo I, Vieira FJ, Roseli D, Melo F, Orientadora DB, et al. Ethnobotanical Survey in the Rural Community Tambaqui. **2011,** *1989*, 30–32.

102. Ferreira, F. M.; Lourenço, F. J. D. C.; Baliza, D. P. Levantamento etnobotânico de plantas medicinais na comunidade quilombola Carreiros, Mercês, Minas Gerais. *Revista Verde de Agroecologia e Desenvolvimento Sustentável* **2014,** *9* (4), 205–212.
103. Guzmán-Gutiérrez, S. L.; Navarrete, A. Pharmacological Exploration of the Sedative Mechanism of Hesperidin Identified as the Active Principle of *Citrus sinensis* Flowers. *Planta medica* **2009,** *75* (04), 295–301.
104. Al-Ashaal, H. A.; El-Sheltawy, S. T. Antioxidant Capacity of Hesperidin from Citrus Peel Using Electron Spin Resonance and Cytotoxic Activity Against Human Carcinoma Cell Lines. *Pharma. Biol.* **2011,** *49* (3), 276–282.
105. Munne, S. L.; Parwate, D. V.; Ingle, V. N.; Tukadoji, R. In Vitro Anthelmintic Activity of Citrus Sinensis Seed Coats. *Int. J. Pharmacol. Res.* **2012,** *2* (2), 83–85.
106. Ali, S.; Prasad, R.; Naime, M.; Zafar, H.; Mahmood, A.; Routray, I. et al. Dried Peel Fraction of *Citrus sinensis* Partially Reverses Pathological Changes in Rat Model of Liver Cirrhosis. *Mediterranean J. Nutr. Metabol.* **2011,** *4* (1), 57–67.
107. Parmar, H. S.; Kar, A. Antidiabetic Potential of *Citrus sinensis* and *Punica granatum* Peel Extracts in Alloxan Treated Male Mice. *Biofactors* **2007,** *31* (1), 17–24.
108. Faturi, C. B.; Leite, J. R.; Alves, P. B.; Canton, A. C.; Teixeira-Silva, F. Anxiolytic-Like Effect of Sweet Orange Aroma in Wistar Rats. *Progress Neuro-Psychopharmacol. Biol. Psych.* **2010,** *34* (4), 605–609.
109. Lehrner, J.; Marwinski, G.; Lehr, S.; Johren, P.; Deecke, L. Ambient Odors of Orange and Lavender Reduce Anxiety and Improve Mood in a Dental Office. *Physiol. Behav.* **2005,** *86* (1–2), 92–95.
110. Goes, T. C.; Antunes, F. D.; Alves, P. B.; Teixeira-Silva, F. Effect of Sweet Orange Aroma on Experimental Anxiety in Humans. *J. Altern. Complement. Med.* **2012,** *18* (8), 798–804.
111. Cappello, A. R.; Dolce, V.; Iacopetta, D.; Martello, M.; Fiorillo, M.; Curcio, R. et al. Bergamot (Citrus bergamia Risso) Flavonoids and Their Potential Benefits in Human Hyperlipidemia and Atherosclerosis: An Overview. *Mini Rev. Med. Chem.* **2016,** *16* (8), 619–629.
112. Cirmi, S.; Bisignano, C.; Mandalari, G.; Navarra, M. Anti-Infective Potential of *Citrus bergamia* Risso et Poiteau (Bergamot) Derivatives: A Systematic Review. *Phytother. Res.* **2016,** *30* (9), 1404–1411.
113. Navarra, M.; Mannucci, C.; Delbò, M.; Calapai, G. *Citrus bergamia* Essential Oil: From Basic Research to Clinical Application. *Front. Pharmacol.* **2015,** *6*, 36.
114. Tranchida, P. Q.; Presti, M. L.; Costa, R.; Dugo, P.; Dugo, G.; Mondello, L. High-Throughput Analysis of Bergamot Essential Oil by Fast Solid-Phase Microextraction–Capillary Gas Chromatography–Flame Ionization Detection. *J. Chromatogr. A* **2006,** *1103* (1), 162–165.
115. Navarra, M.; Ferlazzo, N.; Cirmi, S.; Trapasso, E.; Bramanti, P.; Lombardo, G. E.; Gangemi, S. Effects of Bergamot Essential Oil and Its Extractive Fractions on SH-SY5Y Human Neuroblastoma Cell Growth. *J. Pharm. Pharmacol.* **2015,** *67* (8), 1042–1053.
116. Russo, R.; Ciociaro, A.; Berliocchi, L.; Cassiano, M. G. V.; Rombolà, L.; Ragusa, S.; Corasaniti, M. T. Implication of Limonene and Linalyl Acetate in Cytotoxicity Induced by Bergamot Essential Oil in Human Neuroblastoma Cells. *Fitoterapia* **2013,** *89*, 48–57.

117. Mollace, V.; Sacco, I.; Janda, E.; Malara, C.; Ventrice, D.; Colica, C. et al. Hypolipemic and Hypoglycaemic Activity of Bergamot Polyphenols: From Animal Models to Human Studies. *Fitoterapia* **2011,** *82* (3), 309–316.
118. Impellizzeri, D.; Bruschetta, G.; Di Paola, R.; Ahmad, A.; Campolo, M.; Cuzzocrea, S.; Navarra, M. The Anti-Inflammatory and Antioxidant Effects of Bergamot Juice Extract (BJe) in an Experimental Model of Inflammatory Bowel Disease. *Clin. Nutr.* **2015,** *34* (6), 1146–1154.
119. Risitano, R.; Currò, M.; Cirmi, S.; Ferlazzo, N.; Campiglia, P.; Caccamo, D.; Navarra, M. Flavonoid Fraction of Bergamot Juice Reduces LPS-Induced Inflammatory Response Through SIRT1-Mediated NF-κB Inhibition in THP-1 Monocytes. *PloS One* **2014,** *9* (9), e107431.
120. Braakhuis, A. J.; Campion, P.; Bishop, K. S. Reducing Breast Cancer Recurrence: The Role of Dietary Polyphenolics. *Nutrients* **2016,** *8* (9), 547.
121. Sanguinetti, M.; Posteraro, B.; Romano, L.; Battaglia, F.; Lopizzo, T.; De Carolis, E.; Fadda, G. In Vitro Activity of *Citrus bergamia* (Bergamot) Oil Against Clinical Isolates of Dermatophytes. *J. Antimicrob. Chemother.* **2007,** *59* (2), 305–308.
122. Bagetta, G.; Morrone, L. A.; Rombolà, L.; Amantea, D.; Russo, R.; Berliocchi, L.; Corasaniti, M. T. Neuropharmacology of the Essential Oil of Bergamot. *Fitoterapia* **2010,** *81* (6), 453–461.
123. Saiyudthong, S.; Marsden, C. A. Acute Effects of Bergamot Oil on Anxiety-Related Behaviour and Corticosterone Level in Rats. *Phytother. Res.* **2011,** *25* (6), 858–862.
124. Rangel-Huerta, O. D.; Pastor-Villaescusa, B.; Aguilera, C. M.; Gil, A. A Systematic Review of the Efficacy of Bioactive Compounds in Cardiovascular Disease: Phenolic Compounds. *Nutrients* **2015,** *7* (7), 5177–5216.
125. Berliocchi, L.; Ciociaro, A.; Russo, R.; Cassiano, M. G. V.; Blandini, F.; Rotiroti, D.; Corasaniti, M. T. Toxic Profile of Bergamot Essential Oil on Survival and Proliferation of SH-SY5Y Neuroblastoma Cells. *Food Chem. Toxicol.* **2011,** *49* (11), 2780–2792.
126. Ahmadi, A.; Shadboorestan, A.; Nabavi, S. F.; Setzer, W. N.; Nabavi, S. M. The Role of Hesperidin in Cell Signal Transduction Pathway for the Prevention or Treatment of Cancer. *Curr. Med. Chem.* **2015,** *22* (30), 3462–3471.
127. Celia, C.; Trapasso, E.; Locatelli, M.; Navarra, M.; Ventura, C. A.; Wolfram, J. et al. Anticancer Activity of Liposomal Bergamot Essential Oil (BEO) on Human Neuroblastoma Cells. *Colloids Surfaces B: Biointerfaces* **2013,** *112*, 548–553.
128. Di Donna, L.; Iacopetta, D.; Cappello, A. R.; Gallucci, G.; Martello, E.; Fiorillo, M.; Sindona, G. Hypocholesterolaemic Activity of 3-hydroxy-3-methyl-glutaryl Flavanones Enriched Fraction from Bergamot Fruit (*Citrus bergamia*):"In Vivo" Studies. *J. Functional foods*, *7*, 558–568.
129. Del Río, J. A.; Fuster, M. D.; Gómez, P.; Porras, I.; Garcıa-Lidón, A.; Ortuño, A. Citrus Limon: A Source of Flavonoids of Pharmaceutical Interest. *Food Chem.* **2004,** *84* (3), 457–461.
130. González-Molina, E.; Domínguez-Perles, R.; Moreno, D. A.; García-Viguera, C. Natural Bioactive Compounds of *Citrus limon* for Food and Health. *J. Pharma. Biomed. Analy* **2010,** *51* (2), 327–345.

131. Benavente-Garcia, O.; Castillo, J. Update on Uses and Properties of Citrus Flavonoids: New Findings in Anticancer, Cardiovascular, and Anti-Inflammatory Activity. *J. Agric. Food Chem.* **2008,** *56* (15), 6185–6205.
132. Trovato, A.; Monforte, M. T.; Barbera, R.; Rossitto, A.; Galati, E. M.; Forestieri, A. M. Effects of Fruit Juices of *Citrus sinensis* L. and Citrus limon L. on Experimental Hypercholesterolemia in the Rat. *Phytomedicine* **1996,** *2* (3), 221–227.
133. Bouzenna, H.; Dhibi, S.; Samout, N.; Rjeibi, I.; Talarmin, H.; Elfeki, A.; Hfaiedh, N. The protective Effect of Citrus Limon Essential Oil on Hepatotoxicity and Nephrotoxicity Induced by Aspirin in Rats. *Biomed. Pharmacother.* **2016,** *83*, 1327–1334.
134. Khan, R. A.; Riaz, A. Behavioral Effects of *Citrus limon* in Rats *Metabol. Brain Dis.* **2015,** *30* (2), 589–596.
135. Riaz, A.; Khan, R. A. Behavioral Effects of *Citrus limon* and *Punica granatum* Combinations in Rats. *Metabol. Brain Dis.*, *32* (1), 123–131.
136. Riaz, A.; Khan, R. A.; Algahtani, H. A. Memory Boosting Effect of *Citrus limon*, Pomegranate and Their Combinations. *Pak. J. Pharm. 2014. Sci*, *27* (6), 1837–1840.
137. Dobiaš, L.; Petrová, M.; Vojtko, R.; Kristová, V. Long-Term Treatment with Hesperidin Improves Endothelium-Dependent Vasodilation in Femoral Artery of Spontaneously Hypertensive Rats: The Involvement of NO-Synthase and Kv Channels. *Phytother. Res.* **2016,** *30* (10), 1665–1671.
138. Amorim, J. L.; Simas, D. L. R.; Pinheiro, M. M. G.; Moreno, D. S. A.; Alviano, C. S.; da Silva, A. J. R.; Dias Fernandes, P. Anti-Inflammatory Properties and Chemical Characterization of the Essential Oils of Four Citrus Species. *PloS One*, *11* (4), e0153643.
139. Vaiyapuri, S.; Roweth, H.; Ali, M. S.; Unsworth, A. J.; Stainer, A. R.; Flora, G. D.; Gibbins, J. M. Pharmacological Actions of Nobiletin in the Modulation of Platelet Function. *Br. J. Pharmacol.* **2015,** *172* (16), 4133–4145.
140. Safajou, F.; Shahnazi, M.; Nazemiyeh, H. The Effect of Lemon Inhalation Aromatherapy on Nausea and Vomiting of Pregnancy: A Double-Blinded, Randomized, Controlled Clinical Trial. *Iran. Red Crescent Med. J.* **2014,** *16* (3).

CHAPTER 7

Biological Importance of Essential Oils from the Family Rutaceae

NADIA BIBI[1], MUHAMMAD RIZWAN[2], and MOHAMMAD ALI[2]

[1]*Department of Microbiology, Shaheed Benazir Bhutto Women University, Peshawar, KP, Pakistan*

[2]*Centre for Biotechnology and Microbiology, University of Swat, Swath, KP, Pakistan*

ABSTRACT

Essential oils (EOs) derived from both the upper and subsurface sections of different plants were utilized by Egyptians in burial ceremonies, Greeks associated them to their gods, while Romans used sweet-smelling oils as sedatives. Use of essential oils in modern medicines was started in the 1930s from the concept of aromatherapy of lavender and was practiced as an effective method in wounded soldiers by the French army during World War II. Essential oils are capable of boosting immunity, healing wounds, curing skin irritation, and posing soothing effects on mind and soul. In addition, essential oils can be used to treat diseases of humans, plants, and animals caused by bacteria, fungi, viruses, protozoans, etc. EOs from a variety of plants including family Rutaceae can prevent pathogenic bacterial growth in addition to inhibition of spore germination in fungi. Therapeutic values of essential oils from various members of *Citrus* family have been used to treat fever, gout, rigidity, tenderness, joint pains, abscesses, kidney disorders, and cancers. *Fortunella* species' essential oils have been reported to play a role in the control of *Avian* influenza, while *C. aurantium* and *Z. armatum* have antiprotozoal and antimalarial activities, respectively. Essential oils from

Phytochemical and Pharmacological Investigation of the Family Rutaceae.
Abdur Rauf, Hafiz Ansar Rasul Suleria, & Syed Muhammad Mukarram Shah (Eds.)

the species of *Zanthoxylum* and *Aegles marmelos* have strong analgesic, anti-inflammatory, antioxidant, and antidiabetic activities. Some members of *Citrus* family have a role in the inhibition of biofilm-forming microbes while others have anti-quorum sensing activities.

7.1 INTRODUCTION

Plants have been used for therapeutic purpose including various human ailments for thousands of years, particularly in rustic zones of underdeveloped realms. About 80% of these people relied on the use of customary remedies for healthy life. The consumption of natural products achieved from pharmaceutical herbs and their use in pharmaceutics and agricultural products can be attributed to the presence of a profuse number of bioactive compounds. The use of aromatic plants as a preservative and an enhancer of taste and flavor on one side and for medicine purposes on the other side is very common since ancient times. Prescription of perfume fumigation has already been done by Hippocrates. Use of essential oils was reported in the 16th century for the first time by Paracelsus von Hohenheim, the creator of toxicology, who named these so-called Quinta essential. EOs are composed of natural compounds containing terpenes small molecules composed of carbon, hydrogen, and oxygen. Occurrence of EOs is one of utmost importance in the utilization of plants in the preparation and development of various curative, toiletry, sanitary, and beauty products in addition to their use as food preservatives and additives, agricultural products. Essential oil is for massage purposes and more often used in aromatherapy which is a classical practice.[1]

Essential oils are the complex odorous compounds of volatile nature produced as their secondary metabolites by aromatic plants. The concentrated essence in the form of aromatic essential oils with pleasant fragrance is derived from plants having astonishing therapeutic properties. The aromatic compounds in essential oils are vital for plant survival because these help in pollination by attracting pollinators and preventing them from microbial attacks.[2]

This chapter explores the history of essential oils used in Asian countries from aromatherapy to its uses in modern medicines. The main emphasis is the biological studies of essential oils derived from family Rutaceae as antimicrobial (antibacterial, antifungal, antiviral, antileishmanial, and antiprotozoal), antidiabetic, antiarthritis, analgesics, and anticencerous activities.

FIGURE 7.1 Fruits from citrus family.

Source: Images by left: Ellen Levy Finch (Elf). https://creativecommons.org/licenses/by-sa/3.0/

Right: Nikita from Russian Federation. https://creativecommons.org/licenses/by/2.0/

7.2 HISTORY OF ESSENTIAL OILS

Use of essential oils around the globe is always culture dependent, but it is commonly used for healing purposes at the domestic level. With the passage of time use of essential oils becomes a part of clinical procedures and pharmaceutical products. Farming of Citrus species by prehistoric people is as old as 2100 BC, while the very first recorded history of the aromatic oils can be found in traditional medicines in South Asian countries especially India and China between 3000 and 2000 BC. Use of these EOs in cosmetics and ointments can be traced back as early as ancient Egyptian period around 4500 BC.[3,4] The mixture of different sources of herbal preparations like tuberous plants, resins, cider, carrots, celery and parsley, and grapes in cologne or medication was a common practice in old ages. Information available in Greek ancient times revealed the consumption of different EOs containing marjoram, cumin seeds, rosemary, saffron, and peppermint leaves for the first time between 500 and 400 BC.[5] The role of bioactive compounds caffeine, natural cinchona alkaloid, opioid analgesics, and tropane alkaloid from medicinal plants was identified to show biological activities in the eighteenth and nineteenth centuries. Some EOs such as amaranthine, lavender menthol, and resins are still in use in pharmaceutical products with the hope of best alternatives of unnatural medicines in the future.[6]

7.3 ESSENTIAL OIL SOURCES

EOs can be found in any plant part from bark and stem to roots and leaves to delicate floral parts like petals. However, garden fresh and desiccated seeds and their peels or even resin of most of the plants can be a rich source of these oils like mostly citrus fruits rinds are reservoirs for essential oils. A single essential oil with bioactive compounds containing as many as 100 different chemical components can be derived from different plant genera. Essential oils are used for medical purposes like antimicrobial, antiviral, antioxidants, anticancer, anti-inflammatory, antihistamine, and antidiabetic activities is not new and a well-established mechanism of an antimicrobial action is well implicated nowadays.[1] These natural essences are gifted with strong insecticidal and bioregulatory properties in addition to pest control directly or indirectly and protect stored products. These oils are biodegradable and environment friendly having no toxic effects on aquatic fauna or birds and mammals.[7]

Human sagacity for fragrance is much more than any of other senses and immediately affected by different types of scents. Perfumed chemicals can activate physiological reactions that affect mental and psychological states. These essential oils couple therapeutic properties to renovate stability to the mind, soul, and physique restore balance and vitality, healing, and often an intense sense of well-being. In addition to being pain relieving, anti-inflammatory, antiseptic, and decongesting, oils can comfort anxiety and revitalization of the spirits. EOs play a significant role in attracting insects for the natural breeding of plants through pollination and also help in repulsion of damaging ones.[8]

7.4 AROMATHERAPY

EOs have been reported each with a unique chemistry and assets that are capable of producing a distinct healing, psychosomatic, and biological effect. Today, about 150 essential oils are used in aromatherapy due to their potent ingredients with profound physiological effect, restoring balance and vitality. In addition to being sedative, analgesic, decongesting, and disinfectant, oils can ease anxiety and do revitalization of the moods. Aromatherapy is also known as a complementary part of the phototherapy technique that can be done through spreading the "essential oil in the environment." The process utilizes these oils in flames, vaporizers, and nebulizers to transport aroma through nasal, to induce physical or spiritual effects through breathing or rubbing. The essential oil of *Citrus hystrix* (limau purut or wild lime) is used for aroma therapeutic and pharmaceutical (cosmetics and toiletry products) purposes and antiseptics.[9]

Prevention and curing of many ailments through essential oil is as old as 4000 years ago. Essential oils can act as an alternative medicine due to their harmonizing properties; disorders prevention or cure using essential oils called aromatherapy has been developed.[10]

7.5 ESSENTIAL OILS (EOS) OF FAMILY RUTACEAE

Rutaceae family having strong scented flower placed in order Spindales with 154 genera is commonly known as Citrus family. These species are far spread in the South-East Asia Indo-Malaysia region, India, China, and Pakistan, though its cultivation is universal.[11] Some marketable oil manufacturers make use of different ingredients found in essential oils with diverse aromas in their products. These are biologically active compounds vital to human diet and food, which include vitamins, minerals, flavones, benzopyrene, carbohydrates, and nutritive fibers. Essential oils from Malaysian Citrus plants containing monoterpene hydrocarbon, limonene, are the essential part of many Citrus species such as *C. aurantifolia, C. hystrix,* and *C. microcarpa.*[12] Another species from Rutaceae is *C. grandis* with synonyms *C. decumana* and *C. maxima* is a familiar Malaysian variety famous for its beneficial medicinal properties like its role in fever control, gout, stiffness, inflammation, joint pains, abscesses, and kidney disorders.[9]

FIGURE 7.2 Marketable essential oil from family Rutaceae by manufacturers.

EOs found in members of family Rutaceae for the first time in Pakistan exhibit strong insecticidal and repellent activities. In conventional medicine, the use of *R. graveolens* is for therapeutic properties like gynae problems, skin membrane tenderness, spasms, earache, and headache. EO from Lemon (*Citrus limon)* is capable of boosting up immunity via fabrication of white blood cells, regulation of metabolism and acting as a disinfectant, astringent, and cleansing agent for blemishes associated with oily skin. Many citrus species are known for their curative and pharmacological activities including antibacterials, antifungals, antivirals, antioxidant, anticancer, anti-inflammatory, antidiabetic, insecticidal, antiarthritis, etc.[13]

7.5.1 ANTIMICROBIAL ACTIVITY OF EOS

The antibacterial activities of a variety of extracts from different parts of leaves, fruits, and barks of members of Rutaceae against pathogenic microbes including both bacteria and fungus are shown in Figure 7.3. Similarly, antimicrobial assays of *Ruta graveolens* and *Zingiber officinale* posed an inhibitory ability against *Bacillus cereus*. The antifungal efficacy of many leaf extracts and fractions against causal agents of skin namely *T. rubrum, T. mentagrophytes, Microsporum canis, M. gypseum,* and *E. floccosum* was studied. The antimicrobial activities of EOs derived from *C. aurantifolia* against several uropathogens along with *S. aureus, A. niger,* and *C. albicans* can be due to the presence of bioactive components.[14,15]

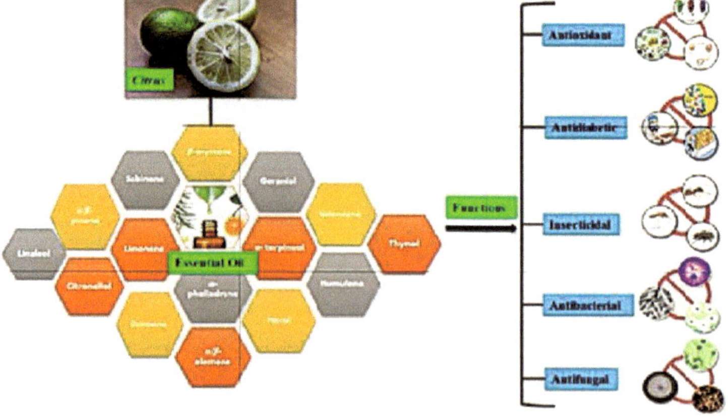

FIGURE 7.3 Citrus essential oil and its various activities.

Source: Reprinted from Ref. [25]. © 2020 by the authors. Licensee MDPI, Basel, Switzerland.https://creativecommons.org/licenses/by/4.0/

7.5.2 ANTIBACTERIAL ACTIVITIES OF EOS

Egyptians used EOs from scented florae for bactericidal activities to prevent food from deterioration. In fact, essential oils derived especially from *Commiphora africana* can act as antibacterial agents against pathogenic bacteria causing typhoid fever (*S. typhimurium)*, sepsis, meningitis, spontaneous abortion, and encephalitis (*L. monocytogenes), dysentery (S. dysenteria), E. coli, B. cereus,* and *oral infection (S. aureus)*. In addition, *C. albicans, H. pylori*, and some Gram-positive and Gram-negative bacteria can also be inhibited by these essential oils. An important bioactivity of EOs is antioxidant activity especially from many bioactivities of *C. hystrix* has been recorded, for example antiseptic, antileukimic, and cough suppressants, while *C. africana* EOs have revealed best inhibitory effects against *H. pylori* activity. EOs from *C. ladaniferus* essential oil were effective against multidrug-resistant strain of *E. aerogenes* and that from *C. hystrix* against *E. coli* and *B. subtilis.*[9]

The EOs obtained from *A. marmelos* showed bioactivity against *B. subtilis, P. aeruginosa S. lutea, A. citreus, E. coli, S. faecalis,* and *S. aureus.*[16] The antibacterial activity of R. graveolens against *Bacillus*, *Enterococcus*, *Staphylococcus*, *Micrococcus*, *Listeria* sps (Gram positive), and *E. coli*, *Klebsiella*, *Pseudomosnas*, *Salmonella*, *Proteus Citrobactor*, *Acintobacter*, and *Enterobactor* (Gram negative) showed a significant antibacterial effect.[17]

EOs of seed and peels of *C. limon*, *C. maxima,* and *C. aurantium* exhibit activity against disease-causing microbes including *E. coli*, *E. faecium P. aeruginosa*, and various *Bacillus species, B. licheniformis, B. altitudinis*, *B. subtilis, S. typhimurium*, *S. aureus*, *S. epidermidis, M. luteus, L. monocytogenese*. Similarly, EOs from *C. reticulata* and *C. sinensis* were found effective against four pathogenic bacteria, *B. subtilis*, and *P. multocida* with strong bactericidal activities.[18] The essential oils derived from *F. margarita* have antimicrobial properties against various pathogenic bacteria causing upper, lower respiratory and urinary tract infection in addition to *B. Subtilis, Staph aureus, S. luta*, *Anthrobacter*, etc.[16]

TABLE 7.1 Antifungal Activities of Essential Oils and Their Components.

S. No.	CEOs	Components	Microorganism	Types of microorganism (bacteria)
1.	Orange, bergamot,		*Arcobacter bulzleri*	Gram –ve
	Lemon EO	Limonene	*E.coli*	
			Klebsiella pneumonia	
			Mycoplasma pneumonia	
			Mycoplasma fermentans	
			Brochotrix thermospacta	Gram +ve
2.	Orange, bergamot		*Arcobacter bulzleri*	Gram –ve
3.	Lemon EO	Linalool	*Staphylococcus epidermidis*	Grame +ve
4.	Orange, bergamot		*Campylobacter jejuni*	Gram –ve
5.	Lemon EO	Citral	*Staphylococcus aureus*	Gram +ve
6.	*Citrus aurantium*	EO	*E. coli*	Gram –ve
			S. typhimurium	
			S. aureus	Gram +ve
			B. cereus	
			L. monocytogenes	
7.	*Citrus limon*	EO	*E. coli*	Gram –ve
			Bacillus subtilis	Gram +ve
8.	*C. reticulate* var. Blanco	EO	*E. coli*	Gram –ve
9.	Orange, bergamot, lemon	EO	*Arcobacter bulzleri*	
10.	Orange, bergamot		*Klebsiella pneumonia*	Gram –ve
11.	Lemon EO		*Staphylococcus aureus*	Gram +ve

Antibacterial activity of *Z. armatum* leaves' essential oil against *E. coli*, *S. aureus*, *P. aeruginosa*, *B. subtilis M. leutus*, *P. multocida,* and *Streptococcus viridines* has been reported from Pakistan with a maximum effect. The north-western Himalayan region of India has also reported antibacterial activity in *Z. armatum* essential oils against the said species.[19]

In addition to *Z. armatum*, *Z. limonella* has been used to treat infectious diseases containing antibacterial and antifungal activities, antituberculous activity against *M. tuberculosis*, Enteropathogens, *S. typhimurium, S. enteritidis, E. coli, S. aureus, C. perfringens,* and *C. jejuni*. EOs from fruits of *Z. limonella* had inhibitory activity for bacterial species *L monocytogenes*, *S. Rissen*, and *P. fluorescens* in addition to *B. cereus, E. coli,* and *S. aureus*. Moreover, the EOs derived from various species of *Zanthoxylum* including *Z. leprieurii, Z. xanthoxyloides, Z. usambarense,* and *Z. chalybeum* also possess strong bactericidal properties for *E. coli, S. aureus.*[20,21]

7.5.3 ANTIFUNGAL ACTIVITY OF EOS

Essential oils from member of Rue are among the most effective and environment-friendly remedy to cope with deleterious effects of fungi, as shown in Table 7.2. Essential oils from *Citrus medica* were found to pose resilient antifungal effects on the growth of pathogenic fungi like *A. niger* and *C. albicans.*[22] The essential oil of *A. marmelos* leaves is proven a remarkable prohibition of *A. niger* and *F. oxysporum* spores at concentrations. The antimicrobial activity of the EOs from *R. graveolens* showed strong inhibitory effects, especially against *C. albicans* in addition to *Fusarium and Alternaria* sp., *oxysporum.*[16]

Moreover, the essential oils of *Z. armatum* have been reported to exhibit diverse biological antifungal properties.[23] Lemon EOs have been shown to have antifungal potential against *Candida* species, namely, *C. tropicalis, C. glabrata,* and *C albicans.*[24]

TABLE 7.2 Antifungal activities of essential oils and their components.

S. No	CEOs	Components	Type of microorganism (fungi)
	Lemon (*C. limon* Burm.f.)	Citral, eugenol	*niger*
	Orange (*C. sinesis* Osb.)	Limonene, citral	*Flavus*
	Orange (*C. sinesis* Osb.)	Limonene, citral	*Fumigatu*
	Pummelo (*C. maxima* Burm.)	Limonene, citral	*Terrreus*
	Lime (*C. hystrix* DC)	Limonene, citronellol, linalool	*Parasiticus*
	Orange (*C. sinesis* Osb.)	Limonene, myrcene	*Penecillium chrysogenum*
	Orange (*C. sinesis* Osb.)	Limonene, myrcene	*P. digitation*
	Mandarin (*C. reticulata* Blanco)	Limonene, ϒ-terpinene	*P. italicum*

Z. limonella fruits' essential oil had inhibitory activity against fungi, especially *Aspergillus ochraceus* and *Fusarium moniliforme* and yeasts such as *R. glutinis*, *S. pombe,* and *H. uvarum.*[20] Essential oils of *F. margarita* have the capability to prevent the growth of *A. niger*, *C. albicans,* and spore germination of *F. oxysporum.*[16]

Food spoilage of fresh and processed food is of utmost importance to the economy of a country. Essential oils from citrus can control molds and other fungal contaminations. These oils possess strong fungicidal activities and help in enhancing shelf life of fruits and vegetables. *C. sinensis* EO was reported to be active for *Aspergillus niger* while that of *Citrus limon* against *Fomitiporia mediterranea* and *Botryospaeria dothidea.* The antifungal activity along with complete inhibition of spore formation of essential oils derived from *C. reticulata* activity against *R. solani, A. alternata C. lunata, F. oxysporum*, and *H. oryzae* were studied. In addition to the essential oils from *C. limon* and *C. sinensis*, two other Citrus species namely *C. aurantium, C. reticulata* showed highest antifungal activity against *Candida albicans* and *Aspergillus flavus.*[25]

7.5.4 ANTIVIRAL ACTIVITIES OF ESSENTIAL OILS

Kumquats (*Fortunella* spp.) member of Citrus family with sweet and edible seed possess a characteristic smell imparted flavonoids and terpenoids. Traditional use of genus *Fortunella* has to treat almost all the disorders from oral to anus tract, especially stomach, liver, gall bladder, and respiratory problems. *F. margarita* and *F. crassifolia* are a rich source of vitamins, flavonoids, carotenoids, and essential oil. A transmissible disease of fowl called *Avian influenza* (AI) is caused by influenza virus Orthomyxoviridae with frequent spread rate. The disease is supposed to be pandemic because of its extraordinary potential for genome level shifts and causes widespread human infection. Extensive spread of highly pathogenic subtype avian influenza (HPAI) in poultry, the H_5N_1 virus, has brought outbreaks in the Asian subcontinent. Therefore, searching for the replacement of antivirals that can effectively restrict influenza viruses including H_5N_1 was plant-derived extracts, especially essential oils. Effects of antiviral properties of EOs from different parts of *F. margarita* against avian influenza (H5N1) virus revealed that the fruits' essential oil was more nominal due to the presence of α-terpineol fruits' EOs.[18]

Citrus oil was applied to H5N1 virus combination with antibiotics on embryonatic eggs chicken erythrocytes based on the potential of allantoic amniotic fluid (AAF) to induce hemagglutination (HA). The essential oils showed a positive hemagglutination test indicates deactivate H_5N_1 virus, but death of the embryos showed the toxicological effect of these oils.[26]

7.5.5 ANTI-PROTOZOAL ACTIVITIES

A major community wellbeing problem in Latin American countries is protozoal diseases, such as amebiasis, giardiasis, trichomoniasis, and trypanosomiasis. The parasitosis of primary concern on the American continent is caused by the protozoa namely, *Trypanosoma cruzi* is Chagas disease. It is a vector-borne disease transmitted to animals and people through insects mostly *Leishmania* species. This disease also can be acquired by humans through transfusions and transplantation of organs and blood congenitally, and oral contamination. Many plants' decoctions and EOs have been used for decades for treatment of several ailments, including parasite infections. EOs from *C. aurantium* and *S. glutinosa* were found best choice for the treatment of Chagas disease in the Tolima region. Moreover, the essential oils *Z. armatum* have also been shown to have antimalarial activity.[27–29]

7.5.6 ANTIARTHRITIS ACTIVITY

G. pentaphylla which is a member of Rutaceae family has been utilized for various ailments in several countries across the globe. It is used to treat rheumatism, stomach, lungs, blood, and liver disorders.[30] Leaves of *A. marmelos, Citrus aurantium,* and *Swinglea glutinosa* were reported to possess antiarthritis activity, and have been extensively used for curative purposes of rheumatism and as antimalarial on account of the presence of numerous bioactive compounds particularly essential oils. The oil extracted from the fruits of *C. aurantifolia* is used for assistance in metabolism and reduction in cholesterol, fats, and sugar level in blood and to treat cold, bronchitis, asthma, allergies, and arthritis.[9]

7.5.7 ANTHELMINTIC ACTIVITIES

EOs derived from different plant parts and fruits of *C. grandis* are among the important constituents of numerous deodorants. *G. pentaphylla,* common name tawshauk, is used in control of different disorders of liver disorders and for anthelminthic activities. Cream of the foliage of this species mixed ginger is applied for eczema and dermal infection and its roots decoction is effective for facial inflammation.[9] The essential oils of *Z. armatum* have been reported to exhibit Antihelmitic activities.[31]

7.5.8 ANTIDIABETIC ACTIVITY

Diverse biological extracts and garden-fresh sap of *A. marmelos* on animal models with streptozotocin and alloxan-induced diabetes were evaluated for the antdiabetic activity. The leaves and callus of *A. marmelos* were found to reduce the blood sugar level due to their antidiabetic potential in streptozotocin induced diabetic rabbits and rats. *A. marmelos* has been reported for its hypoglycemic activities in the traditional medicinal system. The essential oils of *Z. armatum* have been reported to exhibit hepatoprotective properties.[32–34]

7.5.9 EOS AS COX-INHIBITOR

Flesh undeveloped fruits of *A. marmelos* and *Z. armatum* contain bioactive compounds which act as COX inhibitors on carrageenan induce in Inflammation rats causing significant reduction the carrageenan-induced inflammation.[35–37]

7.5.10 ANTIOXIDANT ACTIVITIES OF ESSENTIAL OILS

Reactive-free radicals of ROS-reactive oxygen and RNS nitrogen species are the disruptors of building blocks of body including lipids, proteins, amino acids, and nucleic acid (DNA) through oxidation. These species result in alteration at molecular level during ageing process, cancer, diabetes and asthmatic conditions, arteriosclerosis, an Alzheimer's and Parkinson's disease. Antioxidants are the substances that have free radicals scavenging activity and can protect the cells from oxidative stress (a phenomena of imbalance created free radical and their elimination by the body). Essential oil with the help of its bioactive constituent acts as antioxidants which are capable of bringing balance between free radicals and their rate of removal from the body, for example, Lemon essential oil, which benefits human skin against oxidative damage from skin diseases by regulating the oxidative stress.[10]

Presence of phenolic compounds in EOs derived from *C. limonum* and *F. margarita* acts as antioxidant control in free radical-induced lipid peroxidation and stops dermal tissue destruction. An important bioactivity of EOs is antioxidant activity especially from *C. hystrix, C. aurantifolia* have many bioactivities including antioxidant and anti-inflammatory. Fruits' essential oil of *A. marmelos, Z. limonella,* and *Z. armatum* possesses strong antioxidant and free radical scavenging activity.[14] EOs from peels of sweet orange, *C. sinensis* and *C. limonum*, have robust antioxidant along with membrane stabilization. Inhibition of AChE and (BChE) enzymes by these essential oils can be helpful to cure neurodegenerative problems, particularly Alzheimer's disease. The dried powder of leaves of *Aegle marmelos* can bring about cytoprotection by the stabilization of plasma membrane and variation of antioxidant enzyme machinery of a freshwater fish *(Cyprinus carpio)* on exposure to heavy metals.[25]

7.5.11 INSECTICIDAL PROPERTIES

Essential oils' components not only repel damaging insects but also can be fatal to insects. The renowned target sites on parasites can be structural proteins, enzymes, ion channels, and different transport molecules.[38] Insecticidal properties of EOs from *C. hystrix* and *A. marmelos* against *Tribolium castaneum*, *Sitophilous oryzae*, *Spodoptera litura* (tobacco army worm), and repellant activity against the *S. litura* larvae. EO of *C. hystrix* pose teratogenicity of *C. megacephala*, *C. rufifacies*, *L. cuprina*, and *M. domestica* larvae and toxic effects on adult *C. longa*, *B. rotunda*, and *O. gratissimum* combined with insecticides were studied.[9] In addition, EOs from *C. bergamia* (bergamot) showed insecticidal property against Musa domestica (housefly). Essential oil from *A. marmelos* is capable of stopping attack of pest of stored grains namely, *C. chinensis*, *R. dominica*, *S. oryzae,* and *T. castaneum).* Essential oil when tested on *C. chinensis* significantly reduces egg-lying capacity and adult emergence as well. EOs from the peel of *Citrus* sp. *C. maxima, C. reticulata Blanco, C. suncris, C. sinensis, and C. hystrix* have astonishing insecticidal activities against female *Boophilus microplus* (cattle tick). EO extracted from rind of range was capable of controlling *T. confusum, C. maculatus,* and *S. oryzae*, while that of *C. aurantium* and *C. paradisi* had larvicidal activity against *Anopheles stephensi,* a mosquito vector.[25]

Essential oils of *Citrus sinensis* were demonstrated to be fatal for the larvae and pupae of *Musa domestica*, the housefly displaying highest acaricidal activity. The acaricidal effects on the combination of *C. hystrix* and *C. citratus* on cattle ticks were studied. However, between acaricidal *C. hystrix* and *C. citratus* on *Dermatophagoides pteronyssinus* (house dust mite), the later has higher acaricidal effects toward tick larvae as compared to the former. Mosquito repellent and larvicidal activity of EOs of the fruitlets of *Z. limonella* and *Ruta chalepensis* have been found to show repellent action against *Aedes albopictus mosquito* larvae, *A. aegypti*, *A. dirus*, *C. quinquefasciatus,* and larvicidal against the *Asian tiger* mosquito.[39,20]

7.5.12 ESSENTIAL OILS AS NEMATICIDALS

Nematicidal activity of plant EOs and components from *Gaultheria fragrantissima* and *Zanthoxylum alatum* against the pine wood nematode, *Bursaphelenchus xylophilus* was studied.[37] Essential oils derived from *Citrus bergamia, Citrus limonum, Citrus paradise, Citrus sinensis, Citrus limonum* have resilient Nematicidal activity against *Bursaphelenchus xylophilus*, and the potential of *Haplophyllum tuberculatum* oils to control root knot nematodes has been investigated.[40–41]

7.5.13 ESSENTIAL OILS AS ANTILEISHMANIALS

Leishmanicidal potential of EO from *M. paniculata* leaves has exhibited high Leishmanicidal characteristics when tested against promastigote forms of *L. amazonensis*. Antileishmanial activity contributed to the cytoplasmic membrane disorders, protons dynamic, electric flux, active transport, and clotting of cell matters. Increase in parasite lysis was observed with increase in EO concentration even though EOs from ripe and unripe fruits were active against the parasite. *M. paniculata* inhibited parasite growth regarding leishmanicidal activity which is considered highly active. Antileishmanial activity may be attributed to the mixture of sesquiterpene constituents. The promising leishmanicidal activity exhibited by EO from *M. paniculata* leaves and ripe fruits can be attributed to the chemical constituent *trans*-β-caryophyllene, which was identified at high concentration in EO.[2]

7.5.14 ANALGESIC ACTIVITY

The analgesic activities of methanol extract green foliar parts of *A. marmelos* were screened for pain-relieving by acetic acid-induced writhing test in Swiss mice. Consequently, an enormous reduction in writhing was caused by methanolic extracts, while *C. hystrix* usage in traditional medicines to cure infant's diarrhea, fever, flu, and hypertension, abdominal pains in adults is not new.[9]

7.5.15 ANTICANCER ACTIVITY

Essential oils from *C. aurantifolia* with resilient anticariogenic action can be due to the manifestation of composites with antibacterial potential namely citronellal, citronellol, limonene, and linalool. These constituents of essential oils and foliage and fruit rind of *Citrus aurantifolia* can be inhibitory to the oncogenic bacterial species like Streptococci found in oral cavity, (Streptococcus *oralis*), tooth-decaying *S. mutans*, and *S. salivarius*. Nevertheless, anticencerous effects of EOs from C. *aurantifolia* besides antiaflatoxigenic and anticancer activities proved efficient against cariogenic bacteria.[15]

Methanolic fruit extract of *Z. limonella* comprises effective antitumor promoters or chemopreventive agents as antitumor or antitumor-promoting action showed activation of 12-Otetradecanoylphorbol-13-acetate-induced Epstein-Barr virus early antigen (EBV-EA) in lymphoid cell line. The different extracts fractions of stem barks of *A. marmelos* possess ant proliferative effects against human tumor cell lines including the leukemic K562, TK lymphoid, erythroleukemic HEL, melanoma in addition to breast cancer and MDAMB-231 cell lines can be attributed to the presence of essential oils yet to be discovered.[20]

7.5.16 ANTIDIARRHEAL ACTIVITY OF ESSENTIAL OILS

Medicinal properties of dried fruit pulps *A. marmelos to* control prolonged diarrhea and dysentery confirm the antidiarrheal property of *A. marmelos*. Antidiarrheal activity against the causative organisms of diarrhea against *Shigella boydii, S. sonnei* and *S. flexneri and S. dysenteriae* was studied. Inhibition of *Giardia* and rotavirus was caused by crude water extract of immature fruits of *A. marmelos.*[25]

7.5.17 ESSENTIAL OILS AGAINST QUORUM SENSING

EOs were also tested to check against anti-quorum sensing and were found to inhibit microbes involved namely, *Chromobacterium violaceum* and P. *aeruginosa*. Three species from family Rutaceae *namely Citrus clementina*, *Murraya koenigii*, and Sprengel distinct ingredients possess anti-QS properties that help control food pathogenic mice.[42]

7.6 SUMMARY

Family Rutaceae with universally cultivated members have been used to cure different diseases and their secondary products in the form of essential oils have astonishing characteristic properties from aromatherapy to pharma products and antibacterial to pesticidal and insecticidal activities. This chapter also explores different activities like anthelmintic, antiprozoal, antileishmanial along with antiarthritis and antiviral potentials of these emollients. However, more detailed research is needed to evaluate the anticancer properties, antidiarrheal and analgesic activities of essential oils from family Rutaceae.

KEYWORDS

- **analgesic**
- **anthelminthic insecticidal**
- **antiarthritis**
- **antibacterial activity**
- **anticancer**
- **antidiabetic antioxidants**
- **antiprotozoal**
- **anti-quorum sensing**
- **antitumor**
- **activities**
- **antitussive**
- **history of essential oils**
- **nematicidal**
- **pathogenic microbes**
- **pesticidal**
- **rue family**
- **rutaceae**

REFERENCES

1. Liaqat, I.; Riaz, N.; Saleem, Q. A.; Tahir, H. M.; Arshad, M.; Arshad, N. Toxicological Evaluation of Essential Oils from Some Plants of Rutaceae Family. *Hindawi Evid. Based Complement. Altern. Med.* **2018,** *4394687*, 7.
2. Dhifi, W.; Bellili, S.; Jazi, S.; Bahloul, N.; Mnif, W. Essential Oils' Chemical Characterization and Investigation of Some Biological Activities. *Medicines* **2016,** *3*, 25.
3. Mancini, E.; Camele, I.; Elshafie, H. S. Chemical Composition and Biological Activity of the Essential Oil of Origanum vulgare ssp. hirtum from Different Areas in the Southern Apennines (Italy). *Chem. Biodivers.* **2014,** *11* (4), 639–651.
4. Elshafie, H. S.; Mancini, E.; Camele, I.; Martino, L. D.; Feo, V. D. In Vivo Antifungal Activity of Two Essential Oils from Mediterranean Plants against Postharvest Brown Rot Disease of Peach Fruit. *Ind. Crops Prod.* **2015,** *66*, 11–15.
5. Moore, G. A. Oranges and Lemons: Clues to the Taxonomy of Citrus from Molecular Markers. *Trends Genet.* **2001,** *17*, 536–540.
6. Ali, B.; Al-Wabel, N. A.; Shams, S.; Ahamad, A.; Khan, S. A.; Anwar, F. Essential Oils used in Aromatherapy: A Systemic Review. *Asian Pac. J. Trop. Biomed.* **2015,** *5* (8), 601–611.
7. Maria, J.; Villalobos, P.; Tejero, M. C.; Guirao, P.; Lopez, M. D. Fumigant Toxicity in *Myzus persicae* Sulzer (Hemiptera: Aphididae): Controlled Release of *(E)*-anethole from Microspheres. *Plants* **2020,** *9*, 124.
8. Essential Oils. Mary-Clare Jerram, First American Edition, 2016 Published in the United States by DK Publishing 345 Hudson Street, New York, New York 10014.
9. Othman, S. N. A. M.; Hassan, M. A.; Nahar, L.; Basar, N.; Jamil, S.; Sarker, S. D. Essential Oils from the Malaysian Citrus (Rutaceae) Medicinal Plants Biological Study of Essential Oils from Family Rutaceae. *Medicines* **2016,** *3* (13), 2.
10. Properzi, A.; Angelini, P.; Bertuzzi, G.; Venanzoni, R. Bioactive Essential Oils: Essential Oil as a Source of Bioactive Constituents.) Some Biological Activities of Essential Oils. *Med. Aromat. Plants* **2013,** *2* (5), 136.
11. Mehmood, F.; Manzoor, F.; Khan, Z. U. D.; Ali, M. I.; Khan, I.; Rahim, S. M. A. *Asian J. Chem. Asian J. Chem.* **2012,** *24* (7), 3087–3090.
12. Getaseteg, M.; Tefera, Y. Biological Activities and Valuable Compounds from Five Medicinal Plants. *Nat. Prod. Chem. Res.* **2016,** *4*, 4.
13. Elshafie, H. S.; Camele, I. An Overview of the Biological Effects of Some Mediterranean Essential Oils on Human Health. *Bio.Med. Res. Int.* **2017,** *2017*, 9268468.
14. Sharifi-Rad, M.; Valussi, M.; Tundis, R.; Loizzo, M. R.; Ademiluyi, A. O.; Ayatollahi, S. A.; Iriti, M. Essential Oils: From Plant Chemoecology to Traditional Healing Systems. *Molecules* **2017,** *22*, 70.
15. Lemes, R. S.; Alves, C. C. F.; Estevam, E. B.B.; Santiago, M. B.; Martins, C. H. G.; Dos Santos, T. C. L.; Crotti, A. E. M.; Miranda, M. L. D. Chemical Composition and Antibacterial Activity of Essential Oils from Citrus Aurantifolia Leaves and Fruit Peel against Oral Pathogenic Bacteria. *An Acad. Bras. Cienc.* **2018,** *90* (2), 1285–1292.
16. Ibrahim, N. A.; El-Sakhawy, F. S.; Mohammed, M. M. D.; Farid, M. A.; Abdel-Wahed, N. A. M.; Doaa, A. H.; Rana, B. K. D.; Singh, U. P.; Taneja, V. Antifungal

Activity and Kinetics of Inhibition by Essential Oil Isolated from *Aegle marmelos*. *J. Ethnopharmacol.* **2007,** *57*, 29–34.

17. Reddy, D. N.; Al-Rajab, A. J. Chemical Composition, Antibacterial and Antifungal Activities of *Ruta graveolens L.* Volatile Oils. *Cogent Chem.* **2016,** *2*, 1220055.
18. Chen, Y.; Yang, C.; Wu, Y.; Mo, S.; Wang, S.; Yang, G.; Mei, Z. Glycosmisines A and B: Isolation of Two New Carbazole-indole-Type Dimeric Alkaloids from Glycosmis Pentaphylla and an Evaluation of their Antiproliferative Activities. *Org. Biomol. Chem.* **2005,** *13*, 6773–6781.
19. Singh, A.; Dhami, A.; Palariya, D.; Prakash, O.; Kumar, R.; Kumar, R.; Pant, A. K. Methyl Nonyl Ketone and Linalool Rich Essential Oils from Three Accessions of *Zanthoxylum armatum* (DC.) and their Biological Activities. *Int. J. Herb. Med.* **2019,** *7* (3), 20–28.
20. Supabphol, R.; Tangjitjareonkun, J. Chemical Constituents and Biological Activities of Zanthoxylum limonella (Rutaceae). *Trop. J. Pharm. Res.* **2014,** *13* (12), 2119–2130.
21. Dhami, A.; Singh, A.; Palariya, D.; Kumar, R.; Prakash, O.; Rawat, D. S.; Pant, A. K. α-Pinene Rich Bark Essential Oils of Zanthoxylum armatum DC. from Three Different Altitudes of Uttarakhand, India and their Antioxidant, in vitro Anti-inflammatory and Antibacterial Activity. *TEOP* **2018,** *21* (3), 660–674
22. Chen, C. J.; Li, Q. Q.; Ma, Y. N.; Wang, W.; Cheng, Y. X.; Xu, F. R.; Dong, X. Antifungal Effect of Essential Oils from Five Kinds of Rutaceae Plants – Avoiding Pesticide Residue and Resistance. *Chem Biodivers.* **2019,** *16* (4), e1800688.
23. Guleria, S.; Kumar, A. Antifungal Activity of some Himalayan Medicinal Plants using Direct Bioautography. *J. Cell Mol. Biol.* **2006,** *5*, 95–98.
24. Rashed, A. A.; Rathi, D. N. G.; Nasir, N. A. H. A.; Rahman, A. Z. A. Antifungal Properties of Essential Oils and Their Compounds for Application in Skin Fungal Infections. *Convent. Nonconvent. Approach. Mol.* **2021,** *26*, 1093.
25. Bora, H.; Kamle, M.; Mahato, D. K.; Tiwari, P.; Kumar, P. Citrus Essential Oils (CEOs) and Their Applications in Food: An Overview. *Plants* **2020,** *9*, 357.
26. El-Hawary, S. S.; Ibrahim, N. A.; Mohammed, M. M. D.; Farid, M. A.; Wahed, N. A. M. A.; Ali, M. A.; El-Abd, E. A. W. Antiviral against Avian Influenza (H5N1) Virus and Antimicrobial activities of the Essential Oils of the Leaves and Fruits of Fortunella margarita, Lour. Swingle, Growing in *Egypt J. Appl. Pharm. Sci.* **2015,** *5* (01), 006–012.
27. Tiwary, M.; Naik, S. N.; Tewary, D. K. Chemical Composition and Larvicidal Activities of the Essential Oil of *Zanthoxylum armatum* DC. (Rutaceace) against Three Mosquito Vectors. *J. Vector Borne Dis.* **2007,** *44*, 198–204.
28. Yun, Z.; Ying-Hui, P.; Dong-Qin, Z.; Fei-Fei, C.; Qio-Hui, Q.; Yi, H. Insecticidal Activity of Essential Oil from *Zanthoxylum armatum* Fructification against Two Mosquito Species. *Guihaia* **2012,** *30* (2), 274–279.
29. França Orlanda, J. F.; Nascimento, A. R. Chemical Composition and Antibacterial Activity of *Ruta graveolens* L. (Rutaceae) Volatile Oils, from Sao Luis, Maranhao, Brazil. *South Afr. J. Bot.* **2015,** *99*, 103–106.
30. Sreejith, P. S.; Praseeja, R. J.; Asha, V. V. A Review on the Pharmacology and Phytochemistry of Traditional Medicinal Plant, Glycosmis pentaphylla (Retz.) Correa. *J. Pharm. Res.* **2012,** *5*, 2723–2728.

31. Mehta, M. B.; Kharya, M. D.; Shrivastva, R.; Verma, K. C. Antimicrobial and Antihelmintic Activities of the Essential Oil of *Zanthoxylum alatum* Roxb. *India Perfumer.* **1981,** *25*, 19–21.
32. Ranawat, L. S.; Bhatt, J.; Patel, J. Hepatoprotective Activity of Ethanolic Extracts of Bark of *Zanthoxylum armatum* DC in CCl4 Induced Hepatic Damage in Rats. *J. Ethnopharmacol.* **2010,** *127* (3), 777–780.
33. Qadir, M. I.; Ahmed, B.; Ali, M.; Saleem, M.; Ali, M. Natural Hepatoprotectives: Alternative Medicines for Hepatitis. *RGUHS J. Pharm. Sci.* **2013,** *3* (4), 26–34.
34. Karki, H.; Upadhayay, K.; Pal, H.; Singh, R. Antidiabetic Potential of *Zanthoxylum armatum* Bark Extract on Streptozotocin-Induced Diabetic Rats. *Int. J. Green Pharm.* **2014,** *8*, 77–83.
35. Guo, T.; Denq, Y. X.; Xie, H.; Yao, C. Y.; Cai, C. C.; Pan, S. L.; et al. Antinociceptive and Anti-inflammatory Activities of Ethyl Acetate Fraction from *Zanthoxylum armatum* in Mice. *Phytotherapy* **2010,** *82* (3), 347–351.
36. Sati, S. C.; Sati, M. D.; Raturi, R.; Badoni, P.; Singh, H. Anti-inflammatory and Antioxidant Activities of *Zanthoxylum armatum* Stem Bark. *Global J. Res. Eng.* **2010,** *11* (5), 19–22.
37. Kim, J.; Seo, S.; Park, I. Nematicidal Activity of Plant EOs and Components from *gaultheria fragrantissima* and *Zanthoxylum alatum* against the Pine Wood Nematode, Bursaphelenchus xylophilus. *Nematology* **2011,** *13*, 87–93.
38. Mehmood, F.; Manzoor, F.; Khan, Z. U. D.; Ali, M. I.; Khan, I.; Muhammad, S.; Rahim, A. Evaluation of Toxicity and Repellency of Essential Oils of Family Rutaceae Against Black Ants (*Lasius niger*) in Pakistan. *Asian J. Chem.* 2012, *24* (7), 3087–3090.
39. Conti, B.; Leonardi, M.; Pistelli, L.; Profeti, R.; Ouerghemmi, I.; Benelli, G. Larvicidal and Repellent Activity of Essential Oils from Wild and Cultivated Ruta chalepensis L. (Rutaceae) against Aedes albopictus Skuse (Diptera: Culicidae), An Arbovirus Vector. *Parasitol. Res.* **2013,** *112* (3), 991–999.
40. Park, K.; Park, J. Y.; Kim, K. H.; Choi, K. S.; Choi, I.; Kim, C. S.; Shin, S. C. Nematicidal Activity of Plant Essential Oils and Components from Garlic (Allium sativum) and Cinnamon (Cinnamomum verum) Oils against the Pine Wood Nematode (Bursaphelenchus xylophilus). *Nematology* **2005,** *7* (5), 767–774.
41. Onifade, A. K.; Fatope, M. O.; Deadman, M. L.; Al-Kindy, S. M. Z. Nematicidal Activity of *Haplophyllum tuberculatum* and *Plectranthus cylindraceus* Oils Against *Meloidogyne javanica. Biochem. Syst. Ecol.* **2008,** *36*, 679–683.
42. Camele, I.; Elshafie, H. S.; Caputo, L.; De Feo, V. Anti-quorum Sensing and Antimicrobial Effect of Mediterranean Plant Essential Oils Against Phytopathogenic Bacteria. *Front. Microbiol.* **2019,** *19* (10), 2619.

APPENDIX-A

CHAPTER 7

Biological Importance of Essential Oils from the Family Rutaceae

Nadia Bibi, Department of Microbiology, Shaheed Benazir Bhutto Women University, Peshawar, KP, Pakistan

Muhammad Rizwan, Centre for Biotechnology and Microbiology, University of Swat, Swath, KP, Pakistan

Mohammad Ali, Centre for Biotechnology and Microbiology, University of Swat, Swath, KP, Pakistan

CHAPTER 8

Nutritional Uses of the Family Rutaceae

MUHAMMAD SALEEM[1], MOHAMMAD ALI[2], and
ALLAH BAKHSH GULSHAN[3]

[1]*Department of Chemistry, Ghazi University, Dera Ghazi Khan, Pakistan*

[2]*Institute of Chemical Sciences, Gomal University, Dera Ismail Khan, KPK, Pakistan*

[3]*Department of Botany, Ghazi University, Dera Ghazi Khan, Pakistan*

ABSTRACT

All the historical mysteries remain unsolved, but the cultivation and production of sour-tasted fruits, that is, genus citrus and family Rutaceae, declared as both nutritious and beneficial; that is why it stands on the high score in other fruit crops. Within the past few decades, industrial production and consumption of concentrated citrus fruit juices and pulps have increased considerably. A sour taste, pulpy fruits characterize Rutaceae member, and fibery appearance having a unique mouthwatering aroma which not only feels good but makes the health good by providing specified macro and micronutrients. This study aimed to evaluate the nutritional use of the Rutaceae family. Freshly picked citrus fruits avail a source of vitamin C, Carbohydrates (i.e., glucose and sucrose), a lower proportion of fat. Due to their low-fat content, they are highly beneficial to lowering cholesterol. Vitamin C source is essential to circulate blood, healthy glowing skin, and gastrointestinal disorders. Vitamin C is the most abundant component, while Vitamin B is also present as thiamin, Riboflavin, and niacin. This family

Phytochemical and Pharmacological Investigation of the Family Rutaceae.
Abdur Rauf, Hafiz Ansar Rasul Suleria, & Syed Muhammad Mukarram Shah (Eds.)

can save health from various dangerous disorders and exhibit antioxidant, antimicrobial, antifungal, and even anticancer disorders.

8.1 INTRODUCTION

Family Rutaceae comprises 1500 species and 150 genera spreading in tropical and temperate regions, among which 25 genera and 80 species are reported from India.[1] Rutaceae family members generally include aromatic, pulpy citrus-flavored fruits having sour tastes, for example, grapefruit, lemon, and orange. Members include green, small trees and exhibit a wide hybrid variety with oranges, lime, and grapefruits. Essential oils of Rutaceae are highly useful in nutritional supplements, cosmetic industries, and aromatherapy. Some members' peel is beneficial as dietary supplements containing nonstarch polysaccharides like cellulose. *Citrus reticulata* is a large species with Mandarin fruit, primarily available in China. Mandarin provides 22–25% world's citrus production among commercially available resources.[2]

8.1.1 NUTRITIONAL VALUE OF ORANGE (CITRUS MAXIMA, CITRUS X SINENSIS)

Sour fruit evaluation history of nutrients is related to tropical and subtropical regions of Asia. Family Rutaceae has a significant impact on the economy and has importance in climate and geographical regions because of the nutrients present in specific members of species of family Rutaceae.[3] Genus Zynthoxylum contains 549 species and provides the primary source of amides, alkaloids, terpenes, and flavonoids. Zynthoxylum is rich in nutrients and valuable in dental problems and sexual impotence and stored food products. On extraction, citrus fruits give several active nutrient compounds used in hypertension treatment. Some Rutaceae family members provide antioxidant properties suitable for heart disease treatment. Rutaceae's characteristic properties are antifungal, antibiotic, and anticancer.[4] Rutaceae is a generally well-known citrus family. It comprised blossoming plants with almost 160 genera and contained some flowering classes. Citrus such as lemon (*Citrus limon*), orange (*Citrus sinensis*), grapes (*Citrus paradisi*) has great economic importance. Other citrus fruits contained lime (typically *Citrus aurntifolia*), Fagara or Zanthoxylum, and Agathosma, and some classes of genus Fagara have been found antimicrobial.[5]

FIGURE 8.1 Diagrammatic representation of various species of flowers, fruits, and leaves of family Rutaceae.

Source: Reprinted from Ref. [27]. Copyright © 2010 Elsevier Inc.

Local name: Jambura/Sangtra.

Scientific name: *Citrus maxima, Citrus X sinensis.*

Habitat: Occur in home gardens and in lanes on the side of roads, trees are small in size.

Nutrients: Polyphenol, Folate, Vitamin C, and sugar content.

The seeds are generally considered the waste product of plants, but this suggests that the oil of the *Citrus sinensis* (sweetened orange) has distinctive nutritional importance. That is why this research's primary aim is to explore the nutritious importance of citrus oil.[6] The oil was extracted and purified from the seeds of *Citrus sinensis*, which belongs to the Rutaceae family, and order Sapindales. For that purpose, finely crushed seeds (total 2.0 kg) were soaked in n-Hexane (15 L) for 72 hours and set in with randomly violent trembling, and the extraction of oil was done with soxhlet apparatus. The standard method was used to analyze the chemical constituents of the oil.[7]

The complete analysis showed that the antinutrient level in the oil was within range of safe level. At the same time, mineral and vitamins were recorded in the recommended level of intake limit.[8] The recorded composition of the seed oil of *Citrus sinensis* showed that the oil is rich in: carbohydrate 1.613 ± 0.037, crude fiber 0.618 ± 0.053, ash content 1.786 ± 0.171, fat 90.71 ± 0.292, moisture content 4.736 ± 0.163, and protein 1.560 ± 0.120. The chemical composition of the oil was also recorded by using GC/MS analysis, and this analysis confirms the presence of fatty acid constituents; glutaric acid 2.616%, butyl linoleate 2.867, stearic acid 5.839%, pentadecanoic acid 9.652%, isopropyl linoleate 12.699%, palmitic acid 28.051%, and linoleic acid 37.644%.[9] After conforming to these pharmacologically important constituents present *Citrus sinensis* (sweet orange), their seeds are a superb choice for industrial and nutritional purposes. *Citrus sinensis* is commonly known as Sweet orange, of family Rutaceae; this is a fruit-bearing plant that provides edible things, spread in tropical and subtropical areas. Fruits are recognized as an essential food at the international level.[10] It is said in the reported data that this plant has been growing natively for many years in southeast Asia. However, it is also known from the literature, and these are growing globally. The typical appearance of *Citrus sinensis*, which is also called sweet orange, is closely similar to species like sour orange, *Citrus aurantium*, mandarin orange, and *Citrus reticulate*. Although it is small and evergreen tree and also is about 7.5 m tall, it is also seen that sometimes they are 15 m tall. Its production is approximately one hundred and twenty million (120 million) tons per year at the international level. Therefore, it has great importance in producing fruits among the fruit crops.[11] Around 7.62 million hectares of land are cultivated with various types of Citrus. As a result, the business of citrus fruits has shown tremendous development potential, and it is one of the most efficient agriculture productions in tropical and subtropical regions.[12] The industrial sector processes sweet oranges (*Citrus sinensis*) extensively to produce candies, juices, and pulps. Their extracts are important dietary sources of vitamin C and other natural supplements such as synephrine, limonoids, polyphenols, hesperidin, pectin's, potassium, magnesium, and calcium thiamine and niacin.[13] Furthermore, the waste created by orange processing accounts for around half of the fruits either by industries or by the customer, who throws away the seeds after taking the juice. On an estimated three million hectares of land in Nigeria, around 930,000 tons of citrus fruits are produced each year. However, between 30% and 50% of these citrus fruits deteriorate on their route to the ultimate customers in metropolitan areas. As a result, it is regarded a waste and, as a result, has an environmental impact.[14]

With the goal of repurposing such wastes, reducing environmental impact, and offering a nutritious alternative,. it has great potential in the food application, cosmetic industries, as well as pharmaceutical. It is reported from the previous research that the essential oil isolated from the sweet orange C. sinensis has a variety of activities among which the antifungal activity against L. sajor-caju is more pronounced. It is a type of fungus which causes whiteness rot disease in the wood. Its oil is usually used in diet, because they contain significant amount of biological active compounds with antioxidant activities.[15]

8.1.2 VITAMINS AND MINERALS *IN SWEET ORANGE CITRUS SINENSIS SEED OIL*

Citrus sinensis seed oils contain vitamins and minerals which are shown in Tables 8.1 and 8.2. The research revealed the presence of various minerals and vitamins in the seed oil of *Citrus sinensis* (Sweet orange). The current study revealed that the vitamin A, that is, 0.77 ± 0.012 mg, vitamin K 0.033 ± 0.003 mg, then vitamin E 0.017 ± 0.006 mg, and lowest vitamin D 0.0082 ± 0.0015 mg showed that they are best result in high vitamin's composition found in fixed oil of *Citrus sinensis* as mentioned in Table 8.1. The table contents (Table 8.2) showed the presence of minerals, which are examined. The following minerals were discovered to be present: iron (Fe) 6.99 ± 0.41, calcium (Ca) 15.80 ± 0.10, sodium (Na) 4.60 ± 0.30, manganese (Mn) 0.07 ± 0.01, potassium (K) 4.80 ± 0.20, magnesium (Mg) 6.22 ± 0.22, phosphorus (P) 407.40 ± 2.00, and zinc (Zn) 0.28 ± 0.02.[16]

TABLE 8.1 Composition of Different Vitamins in Sweet Orange *Citrus sinensis* Seed Oil.

Vitamins	Values (mg) ± SEM	DSID unit equivalent (IU and μg)
Vita A	0.77 ± 0.012	2533.34 IU
Vita D	0.0085 ± 0.0015	337 IU
Vita E	0.017 ± 0.006	23.89 IU
Vita K	0.033 ± 0.003	33 μg

± indicates that the experiment was performed in triplicate. It is SEM.

TABLE 8.2 Content of Various Minerals in Seed Oil of Orange *Citrus sinensis.*

Name of mineral	Value in (mg/L)	FDA (recommended daily limits) (mg)
Iron (Fe)	6.99 ± 0.42	18
Calcium (Ca)	15.80 ± 0.12	1000
Sodium (Na)	4.60 ± 0.31	2400
Manganese (Mn)	0.07 ± 0.02	2
Potassium (K)	4.80 ± 0.23	3500
Magnesium (Mg)	6.22 ± 0.23	400
Phosphorus (P)	407.49 ± 2.03	1000
Copper (Cu)	± 0.0001	2

8.1.3 PROXIMATE

For parameter assay, the direct transcription of *Citrus sinensis* seed oil was examined and experiment was performed in triplicate. The results are shown in Table 8.3. As the result following is the quantities of the proximate components given in the percentage, that is, fats 89.812 ± 0.283, proteins 1.562 ± 0.122, ash contents 1.776 ± 0.161, crude fibers 0.62 ± 0.053, moisture contents 4.645 ± 0.154, and carbohydrates 1.615 ± 0.035.[17]

TABLE 8.3 Analysis Results of *Citrus sinensis* Seed Oil with Proximate Composition.

Parameters	Content (%)
Protein	1.562 ± 0.122
Ash content	1.776 ± 0.161
Crude fiber	0.62 ± 0.053
Moisture content	4.645 ± 0.154
Fat	89.812 ± 0.283
Carbohydrate	± 0.035

8.2 BITTER-ORANGE

Roadsides and farmhouse gardens are ideal habitats for this small- to medium-sized tree. From February through November, the plant blooms and bears fruits. Southeast Asia is where this species is thought to have originated. It is now grown across the tropics. It is planted in almost every part of Bangladesh for its delicious, luscious fruits.[18] It is a small- to medium-sized spinous tree, according to the taxonomic classification. Petioles are widely winged, and the leaflets are big, 15–23 cm long, ovate-oblong, and ovate-oblong. It has large, light yellow, globose, or pyriform fruits with thick skin and red to pale pink or yellow flesh. Conventional Therapeutic Uses: the fruit is nutritious, cardiac tonic, and refrigerant and is beneficial in treating influenza, cough, and asthma. The rind is an anthelmintic that can help with vomiting, abdominal cramps, and diarrhea. The leaves are beneficial in treating epilepsy and convulsive cough.[20]

8.2.1 TOOTH-BRUSH PLANT

Local name: Kaghzi limbu.

Scientific name: *Citrus aurantifolia.*

Habitat: Tree is small in length which exists on dry land and in gardens.

Nutrients: Sugar, lipids, phenols, and nitrogenous compounds are all present in the juice. Maleic acid, succinic acid, and citric acid are abundant. Lipoid, phenols, and nitrogenous compounds are all found in juice.

Acid lime (*Citrus aurantifolia*) is a member of the Citrus genus in the Rutaceae family, with chromosome number 2n=18. Its origins may be traced back to India. In India, it is popularly referred to as nimbu. Nimbu is also known as sour lime, acid lime, Indian lime, key lime, and other similar terms.[21] Gujarat, Andhra Pradesh, Maharashtra, Odisha, and Madhya Pradesh are the leading states in India in terms of lime/lemon area and production. The area under lime/lemon is 296 thousand hectares, with an output of 3397 thousand MT and productivity of 11.47 MT/Ha.

8.2.2 NUTRITIONAL AND BIOCHEMICAL COMPOSITION

Acid lime juice is naturally acidic, with high citric acid, malic acid, and succinic acid levels. The primary vitamins found in the apple are vitamin

C and B. Chlorophyll and carotenoids are pigments found in the Flavedo of fruit. Cellulose, hemicellulose, lignin, and pectin are abundant in albedo. Oil glands in Flavedo are used to extract essential oils for use in the cosmetics and perfume industries. Limonene, citronellol, neral, γ -terpinene, and β-pinene are the most common volatile chemicals found in fruit peel oil. Lime juice contains sugars, lipids, nitrogenous chemicals, and phenols. Flavones, flavanols, and flavanones are all important flavonoids in the diet.[22]

8.2.3 MEDICINAL PROPERTIES

The vitamin C content of the acid lime fruit is high, which aids in disease resistance. Using the fruit reduces toothache, dental cavities, and sore gums. Lime aids digestion and aids in the reduction of acidity, constipation, and peptic ulcers. It will also aid in the treatment of obesity. The fruit's vitamin C content will help to heal scurvy. It treats respiratory problems, urinary problems, piles, rashes on the skin, and dandruff. Lime has anticancer, anti-rheumatoid arthritis, antidiabetic effects, and helps treat cardiovascular disorders.[23]

8.3 GREEN APPLE AEGLE MARMELOS CORR.

Local name: Bel, *Feronia Lemonia*

Scientific name: *Aegle marmelos Corr.*

Habitat: Medium-sized length which exist in muddy soil as well as in dry lands.

Nutrients: Vitamin A, B1, B2 that other fruits lack and these are declared helpful in asthma, dysentery, hepatitis, and tumor.

Nutrition-rich vegetables and fruits have been shown to reduce the risk of chronic diseases such as cardiovascular disease and cancers. Phytochemicals present in fruits and vegetables, such as phenolics, flavonoids, and carotenoids, may play a key role in decreasing chronic disease risk. The epidemiological and phytochemical investigation of a typical fruit apple has great nutritious importance, revealing that apple consumption decreases the risk of some cancers, heart disease, asthma, and diabetes. Apples have been proven to have significant antioxidant activity in laboratory tests and inhibit cancer cell development, reduce lipid oxidation, and decrease cholesterol. Apples are rich in various phytochemical constituents, containing catechin, quercetin, chlorogenic acid, and phloridzin, all high oxidant activity. Although there are slight differences in phytochemical contents as the fruit grows and ripens, the concentration of phytochemical constituents in apples differs dramatically among varieties. The phytochemical contents of apples are little to not affected by storage, but they are substantially affected by processing. While there is a lot of data out there, there has not been a comprehensive assessment of the medical importance of the phytochemical contents of apples. The goal of the current study is to examine the latest research on the nutritional importance of apples because of their phytochemicals, bioavailability of phytochemical and antioxidant activities, and the impacts of a range of storage, ripening, and processing on apple chemical content.[24]

TABLE 8.4 Contents of Various Nutrients in Green Apple *Aegle Marmelos Corr.*

Composition	Amount
Total phenol	236 mg/100 g
Pectin%	1.38–1.66
Vitamin C	1.69–3.41 mg/100
Na%	3.08–7.53
Titratable acidity	0.84–2.77
Total carbohydrates	15.7 g/100 g
TSS (Triterpinoids)	9.41–16.02
Protein	6.35 g/100 g
Total sugar	–4.48%

8.4 WOOD APPLE

Botanical name: Aegle marmelos (L.) Corr.

Local name: Sharbati limoo.

Scientific name: *Citrus limonum, Citrus limoo.*

Habitat: Small tree which grows in homes, gardens, and fields.

Nutrients: Major nutrient is Vitamin C along with B1 and B2.

8.4.1 INDUSTRIAL PROCESSING OF CITRUS AND ITS BY-PRODUCTS

Citrus belonging to the Rutaceae family is one of the world's most significant fruit crops, growing in tropic and subtropics regions across the world and regarded for its worldwide accessibility and attractiveness. Due to the complicated biochemistry comprehending the existence of apomictic clones, hybrid lines, and the broad geographical dispersal allowing cross-species crossing, the taxonomic classification of the genus citrus has yet to be adequately defined, creating a continual issue for botanists. The most widely used industrialized fruit products belong to the following species: *C. reticulata* (tangerine), *C. Sinensis* (sweet orange), *C. Limon* (lemon), *C. aurantifolia* (lime), and *C. paradisi* (grapefruit). Global citrus fruit output was relatively constant between 2008 and 2016, fluctuating between a minimum of 115,541.8 tons (2009) and a maximum of 132,002.3 tons (2012).[25] In 2016, it was around 124,246.0 tons, with the Northern Hemisphere (97,848.9 tons) accounting for more than the Southern Hemisphere (97,848.9 tons) (26,397.1 tons). In the boreal hemisphere, China (32,705.9 tons) and the Mediterranean area (25,216.0 tons) dominated output, while Brazil (16,555.1 tons) topped the austral hemisphere. In the same year, sweet oranges (66,974.1 tonnes) were confirmed to be the most produced fruit, followed by tangerines (32,968.5 kg), lemons, limes (15,981.8 kg), and grapefruit (15,981.8 thousand tonnes) (8321.6 tonnes). The fresh fruits of citrus were formerly sold exclusively and consumed as fresh fruits due to exceptional postharvest durability that facilitated global trade. However, as time has gone, new financial implications have evolved, and fruit processing has become a need to meet customers' changing needs and keep farmers' businesses afloat, particularly at the local level. As a result, due to small- and medium-sized enterprises worldwide, the generation of liqueurs, marmalades, essential oils, jams, and canned citrus fruits has grown in popularity. On the other hand, large-scale manufacturing began in the early 1900s in selected US states such as Florida and California, establishing citrus juice industries. In 2008–2016, the volume of fruit destined for processing fluctuated between 29,400.0 thousand tons and 23,538.9 thousand tons, representing 25% and 20% of total citrus fruit output, respectively. The sweet orange was the most processed fruit type globally in 2016 (18,460.9 tons), followed by lemons (1821.5 tones), limes (787.5 tons), tangerines (2469.0 tons), and grapefruits (2469.0 tons). As a result of the enormous quantity of citrus fruits processed, a large amount of garbage was generated, accounting for nearly half of the raw citrus fruits. Citrus trash comprises solid and semisolid wastes (fresh peel, membranes, pulp parts, dried peel and pulp, seeds, sludge, and citrus meal and fines). Citrus waste also represents liquids (typically, citrus sludge from liquid wastes from citrus processing factories) and distillery effluents (effluents from citric acid and pectin production and citrus peel liquor, essential oil plants, and molasses). In addition to the high expenses associated with trash treatment, handling these materials frequently results in regulatory restrictions (e.g., mandated waste generation reduction). Traditionally, citrus waste was handled in the following ways: (1) straight outflow of industrial effluent into lakes; (2) dumps in wastes lakes; (3) dumps into groves/wells; and (4) dumps in drainage systems (e.g., anaerobic, composting, thermolysis, digestion, incineration, and gasification). These approaches might result in serious contamination and, most importantly, vital nutrients and biomass release. As a result, comparable to several food wastes in reducing environmental degradation and building sustainable cyclical economies, in recent decades, citrus by-products have shown to have exceptional recovery potential. Valorizing citrus by-products might be a way to grow the citrus industry or even start a new, related, and frequently separate business. One of the first by-product recovery facilities was created in Ontario (California) in 1927 due to the large amounts of lemons and oranges rejected as cull fruit and had to be destroyed at the cost of at least one $1 per ton. Also, at a time, an industry processed 10 thousand tons' fruit prepared around 50 thousand pounds of essential oil extracted from the peels of citrus that can also be improved, with an average cost of USD100,000. In this context, the purpose of this chapter is to provide a comprehensive overview of the functional and nutritional properties, extraction techniques, and potential implementation of a valuable product derived from the unit of citrus processing, particularly seed oil obtained from lemon (*C. Limon*) by the cold-pressed process. The review's information synthesis may aid in promoting a greener, better eco-friendly usage of lemon seeds oil obtained in a cold-pressed way having better nutritional properties.[26]

8.5 SUMMARY

It is confirmed through the above discussion that green apples bear fruit from 12 to 70 years, full of pectin, protein, vitamin A, B1, B2, due to which it exhibits vast potential for thirst satisfaction. Green apple has a blended sour and sweet taste due to several phytochemicals like tyramine derivatives, coumarins, and saponins. The study confirmed that Orange contains sugar content and phenol, dietary fiber, vitamin C, and folate. Lemon fruit among Rutaceae is a combination of acid juices with a lot of nutritional benefits. It has citric, acetic, and melic acid components in it. It is very useful for gastric problems and indigestion. It improves appetite and makes a person feel fresh. Peel of lemon has cellulose, hemicellulose, pectin, and chlorophyll. It works as an antioxidant and digestion agent used commercially in the medical industry. Thus, the overall view of the Rutaceae family makes it an essential and versatile beneficiary for use based on its nutritional values.

ACKNOWLEDGMENTS

The corresponding authors highly acknowledge the vice chancellor Ghazi University for wonderful environment of academic and research activities in the varsity.

KEYWORDS

- **antioxidant**
- **antimicrobial**
- **antifungal**
- **anticancer**
- **cholesterol**
- **family Rutaceae**
- **glucose and sucrose**
- **niacin**
- **riboflavin**
- **thiamin**
- **vitamin C**

REFERENCES

1. Sharma, O. P. *Plant Taxonomy*; Tata McGraw-Hill Publishing Company Limited, New Delhi, India, 2004; vol 203, pp 111–128.
2. Kondo, M.; Goto, M.; Kodama, A. Fractional Extraction by Supercrit-ical Carbon Dioxide for the Deterpenation of Bergamot Oil. *Ind. Eng.* **2010,** *39*, 474–478.
3. Elleuch, M.; et al. Dietary Fiber and Fiber Rich by-Products of Food Processing. *Food Chem.* **2011,** *4*, 411–421.
4. Tolkowsky, S. In *Hesperides: A History of the Culture and use of Citrus Fruits*; John Bale, Sons & Curnow Ltd, 2010; vol 98, pp 222–234.
5. Gmitter, G. Citrus Species (Rutaceae). *Econ Bot.* **2000,** *44*, 267–277.
6. Singh, H. P.; Chadha, K. L.. Genetic Resources of Citrus. *Adv. Hortic.* **2003,** 95–121.
7. Udo, I. Potentials of *Zanthoxylum zanthoxyloides* for the Control of Stored Products Insect Pest. *J. Stored Prod. Post Harvest Res.* **2011,** *2* (3), 40–44.
8. Pragasam, S. J.; Rasool, M. Dietary Component p-coumaric Acid Suppresses Monosodium Urate Crystal-Induced Inflammation in Rats. *Inflamm. Res.* **2000,** *62* (5), 489–498.
9. Waleed, F. Citrus Varieties in Egypt: An Impression. *Int. Res. J. Appl. Sci.* **2019,** *1*, 63–66.
10. Kumar, V.; Subramanian, B. Chromosome Atlas of Flowering Plants of the Indian Subcontinent. *J. Bot. Sur.* **1986,** *33*, 441–480.
11. Ahmed, Z. U.; Abdul Hassan, M.; et al. In *Encyclopedia of Flora and Fauna of Bangladesh*; Asiatic Society of Bangladesh: Dhaka, 2009; vol 10, pp 159–187.
12. Yusuf, M.; Begum, J.; Hoque, M. N.; Choudhury, J. U. *Medicinal Plants of Bangladesh-Revised and Enlarged*; Bangladesh Council of Scientific and Industrial Research, 2009; vol 22, pp 302–444.
13. Prain, D. Bengal Plants. *Bot. Surv. India* **2003,** *22*, 999–1031.
14. Mahbubur Rahman, A. H. M.; et al. A Preliminary Assessment of Angiosperm Flora of Bangladesh Police Academy. *Res. Plant Sci.* **2014,** *2*, 9–15.
15. Uddin, M. Z.; Alam, M. F.; Rahman, A.; Hassan, M. A. Diversity in Angiosperm Flora of Teknaf Wildlife. *J. Plant Taxon.* **2013,** *20*, 145–162.
16. Nithya, N.; Saraswathi, U. In Vitro Antioxidant and Antibacterial Efficacy of *Feronia elephantum* Correa Fruit. *Indian J. Nat. Prod. Resour.* **2014**, *1*, 301–305.
17. Anonymous. In *Indian Horticulture Data Base*; National Horticulture Board, 2018; vol 86, pp 777–790.
18. Ganguly, S. Medicinal Properties of Lime and its Traditional Food Value. *Res. J. Pharm. Sci.* **2013,** *2* (4), 19–20.
19. Nithya, N.; Saraswathi, U. In Vitro Antioxidant and Antibacterial Efficacy of *Feronia elephantum* Correa Fruit. *Indian J. Nat. Prod. Res.* **2010,** *1* (3), 301–305.
20. AIJN. *Liduid Fruit Market Report*; AIJN European Fruit Juice Association, Brussels, 2014.
21. Borah, B. K.; Johnson, A. M.; Sai Gopal, D. V.; Dasgupta, I. Sequencing and Computational Analysis of Complete Genome Sequences of Citrus Yellow Mosaic Badna Virus from Acid Lime and Pummelo. *Virus Genes* **2009,** *39* (1), 137–140.
22. Adefegha, S. A. Functional Foods and Nutraceuticals as Dietary Intervention in Chronic Diseases; Novel Perspectives for Health Promotion and Disease Prevention. *J. Diet. Suppl.* **2018,** *15* (6), 977–1009.

23. Barros, H. R.; Ferreira, T. A.; Genovese, M. I. Antioxidant Capacity and Mineral Content of Pulp and Peel from Commercial Cultivars of Citrus from Brazil. *Food Chem.* **2012,** *134* (4), 1892–1898.
24. Maity, P.; Hansda, D.; Bandyopadhyay, U.; Mishra, D. K. Biological Activities of Crude Extracts and Chemical Constituents of Bael, Aegle marmelos (L.) Corr. *Indian J. Exp. Biol.* **2009,** *47* (11), 849–861.
25. Djenane, D. Chemical Profile, Antibacterial and Antioxidant Activity of Algerian Citrus Essential Oils and Their Application in Sardina pilchardus. *Foods* **2015,** *4* (2), 208–228.
26. Luzia, D. M.; Jorge, N. Oxidative Stability and Alpha-Tocopherol Retention in Soybean Oil with Lemon Seed Extract (Citrus Limon) under Thermoxidation. *Nat. Prod. Commun.* **2009,** *4* (11), 1553–1556.
27. Takhtadzhian, Armen Leonovich. Diversity and classification of flowering plants. Columbia University Press, 1997.

CHAPTER 9

Rutaceae: An Insight into Healthcare and Clinical Applications

ABHAY PRAKASH MISHRA[1], LUBNA AZMI[2], MANISHA NIGAM[3], RAQUEL PERES MORAIS URANO[4], ARCHANA YADAV[5], and MOTLALEPULA GILBERT MATSABISA[6]

[1]*Department of Pharmacology, University of Free State, Bloemfontein, Free State, South Africa*

[2]*Institute of Pharmaceutical Sciences, University of Lucknow, Lucknow, Uttar Pradesh, India*

[3]*Department of Biochemistry, Hemvati Nandan Bahuguna Garhwal University, Srinagar, Garhwal, Uttarakhand, India*

[4]*Instituto de Quimica de Sao Carlos, Universidade de Sao Paulo, Sao Carlos, Brazil*

[5]*Department of Microbiology, Institute of Biological Sciences and Biotechnology, C. S. J. M. University, Kanpur, Uttar Pradesh, India*

[6]*Department of Pharmacology, University of Free State, Bloemfontein, Free State, South Africa*

ABSTRACT

The Rutaceae family members are widely farmed across the world because of their numerous health advantages for humans and the pharmaceutical and food industries. Citrus is the largest genus of the Rutaceae family. Known throughout the world for their several juicy citrus fruits, *Rutaceous*

Phytochemical and Pharmacological Investigation of the Family Rutaceae.
Abdur Rauf, Hafiz Ansar Rasul Suleria, & Syed Muhammad Mukarram Shah (Eds.)

members have a high percentage of vitamin C as well as several alkaloids. The distinctive and active components of Rutaceae are alkaloids, coumarins, flavonoids, limonoids, and volatile oils. These natural products exhibit a variety of biological activities, including anxiolytic, anti-inflammatory, hypoglycemic, anthelmintic, anticancer, and anti-infective activities. Over the past few years, these compounds have attracted tremendous interest and may have the potential to become new drug lead compounds. In this chapter, we have compiled the data available so far on the clinical application of the family Rutaceae.

9.1 INTRODUCTION

Rutaceae, the flowering plant family, comes from the Sapindale order, which is also known as the rue or citrus family. Several prominent edible fruits are produced by the *Citrus* species of the Rutaceae. The plants are extensively spread in China, India, and southeast Asia, but they are also grown in other parts of the world.[12] The leaves and fruits of the Rutaceae family plants comprise a range of bioactive compounds as well as physiologically active chemicals such as folic acid, potassium, vitamin C, coumarins, flavonoids, dietary fibers, and pectin that are vital to human nutrition and diet.[17] The Rutaceae family, which includes approximately 1500 species and 150 genera, is found in both tropical and temperate climates, particularly in Australia and South Africa. So far, India has recorded over 25 genera and 80 species of this family.[25]

Rutaceous members have a high proportion of vitamin C as well as many alkaloids, and are known across the world for several delicious citrus fruits such as our modern-day oranges, lemons, grapefruits, and so on. *Fagara*, *Zanthoxylum*, *Ruta*, *Glycosmis*, *Eriostemon*, *Atalantia*, *Citrus*, *Murraya*, *Aegle*, mandarins (*Citrus reticulata* Blanco), limes (*Citrus latifolia*), grapefruits (*Citrus paradisi*), clementines (*Citrus clementina*), lemons (*Citrus limon*), oranges (*Citrus sinensis*), tangerines (*Citrus tangerine* Hort. Ex Tanaka), and pomelos (*Citrus maxima* Burm. Merr.) are the most well-known of them.[25] Citrus fruits are high in flavonoids, which have been linked to hypolipidemic, hypoglycemic, anti-inflammatory, and antioxidant effects. *Rutaceous* species have a greater overall amount of vitamin C, phenols, flavonoids, and antioxidant constituents. In comparison to lemons (49.8 mg GAE/g) and oranges (35.6 mg GAE/g), grapefruits have the highest concentration of total phenols

(77.3 mg GAE/g).[28] This family is widely used to treat pain, dermatitis, rheumatism, and other inflammatory diseases like CVS and CNS dysfunctions. It is also used as a fragrance, cuisine, and traditional medicine. Various medicinal benefits associated with the family Rutaceae are shown in Figure 9.1. Alkaloids, coumarins, flavonoids, limonoids, and volatile oils are included in this family's phytochemistry.[28] The family's abundant edible fruits of the genus *Citrus* are of significant economic value in temperate and subtropical areas.

Rutaceae has seven subfamilies, *Aurantioideae* being one of them. In the *Micromelinae* subtribe, *Micromelum* is the sole genus, while *Clauseneae* is a tribe of the *Aurantioideae* subfamily. The genus *Micromelum* has yielded coumarins, alkaloids, and flavonoids.[34] There is a broad range of literature about the morphology of the Rutaceae family, but there is not any good explanation of the clinical application of the Rutaceae family medicinal plants. This chapter comprises a detailed explanation of the pharmacological and clinical properties of Rutaceae.

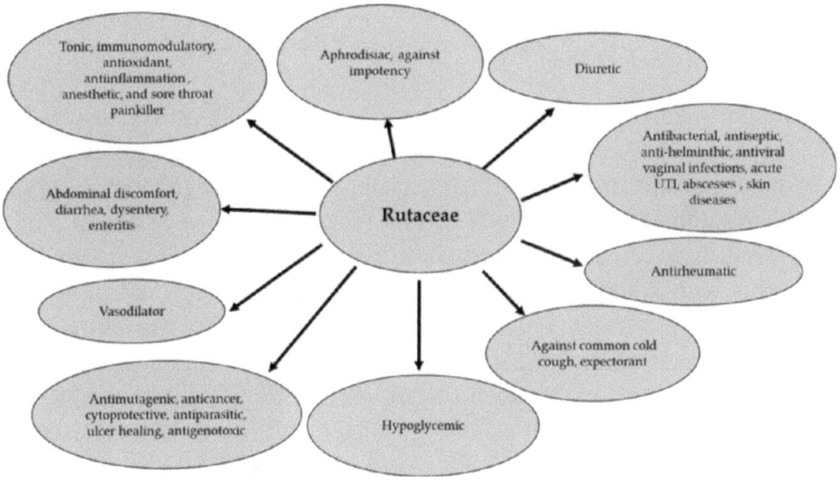

FIGURE 9.1 Medicinal benefits associated with the family Rutaceae.

9.2 CLINICAL APPROACH OF RUTACEAE AS A MEDICINE

Citrus essential oils' antibacterial activity was shown to be stronger than that of phenol as early as 1948.[15] Some of the component molecules that are effective against bacteria, fungus, protozoa, and insects are phytoalexins, which the plant produces in reaction to an attack. Among citrus fruits' peels and other plants, there is a chemical called limonene. Limonene is used to flavor chewing gum, drinks, and foods, and in pharmaceuticals to aid in the penetration of therapeutic ointments and lotions. Different stereoisomers of limonene are responsible for the distinct flavors of orange and lemon juice. Limonene has a variety of biological functions. Differences in the effectiveness of citrus fruits might be attributable to this, at least in part. It's fascinating to note that the citrange, *C. insitorum*, a hybrid between *C. trifoliata* from northern China and an orange, proved to be the most effective. These chemicals are part of the long-lived fruits' natural defenses since they cling to the trees for months. Citral, a combination of two geometric isomeric aldehydes, is one of them and has been overexploited for a long time. In England, a solution made from boiling lemon peel is still used as a home treatment for acne. *C. hystrix* is most known for its citronellal-rich leaves, which are used in cooking, but it is also used to wash hair and to keep leeches away in Sri Lanka, presumably due to its insecticidal chemicals' broad spectrum of effects on invertebrates.[29]

9.2.1 WEIGHT LOSS

Some physicians believed that lemon juice might help people lose weight in the early 20th century, and the grapefruit diet became popular in the 1970s, especially among North American women. The function of grapefruit, a mandarin–pomelo hybrid, has recently been reexamined, and it has been discovered that the fruit's membrane contains a compound that interacts with the liver to decrease cholesterol and assist control insulin.[9] Despite this, a 3-months research found that people who ate half a grapefruit with each normal meal (three times a day) dropped 1.6 kg on an average, compared to 0.2 kg for those on an otherwise comparable diet, with some losing as much as 4.5 kg; juice alone had a detectable impact as well. The study included 1000 obese people who were shown to have lower insulin and glucose levels at the end of the trial.[9]

9.2.2 AIDS

It is determined that citric acid alters surface proteins in the virus's envelope, preventing binding and subsequent infection. Infection due to HIV virus appears to be suppressed due to citric acid at a dose of 5%

and above. Within 2 min, a 50% solution (pH 2.7) containing citric acid killed all viruses.[10] Lemon juice is more effective than lime juice, probably reflecting the lemon's heritage, yet both are better than commercially supplied microbicides. Experiments with macaques have demonstrated that it is safe for primates to use as a vaginal microbicide, and clinical trials are now underway in Thailand. At 30% coverage of the sexually active population, a microbicide with just 40% effectiveness against HIV transmission would prevent more than 5.6 million HIV infections per year; at only 20% coverage, 4 million would be prevented. It's worth noting that human ejaculate already includes citric acid, suggesting that new uses might just be reinforcing a natural microbicide. These discoveries in traditional medicine have clear implications for population control and HIV transmission.[27] There are even more besides pesticides and fungicides. Flavonoids, notably flavonones from the hesperidia, have been extensively investigated for their antioxidant, anticancer, antiviral, and anti-inflammatory properties, as well as for their impact on capillary fragility and suppression of human platelet aggregation. Anti-sweetening agents such as hesperidin and naringin, as well as their byproducts, are commercially valuable flavonoid-based industrial products. And, in the Hindu festival of Thaipusam, devotees hang fruits of one type of *Citrus hystrix* on hooks in self-mortification of the back; whether it's the white ash or milk used at the same time or compounds from the fruits that help to numb the skin is still unknown, though the fruits alone are sometimes used, suggesting they may contain an effective astringent.[27]

9.2.3 CONTRACEPTION

Lemons might be used to identify venereal infection in women, as if the labia are raw due to infection, they would react to the juice. The juice from a small lemon might be used to produce an effective cervical cap. Recent studies have shown that 20% fresh lemon juice in human ejaculate causes 100% of sperm to become irreversibly immobilized in less than 30 s.[27] Citric acid damages mitochondrial proteins, resulting in immobility.[15]

9.2.4 SCURVY

The deadly illness induced by a deficiency of vitamin C, found in fresh fruits and vegetables, used to kill more sailors than their human foes, and a 50% fatality rate was usual among sailing crews until the 19th century.[23] The use of lemon juice to treat the illness dates back to the 17th century. The ship with lemon juice on board was the only one free of scurvy when the East India Company (EIC) dispatched its maiden ships to the east in 1601. After that, all EIC ships carried it, as did those of the Dutch East India Company (VOC) as early as the 1630s, but it had gone out of favor by the end of the century due to the trendy "imbalance of body humors" kind of medicine. The controlled experiment of surgeon James Lind (1716–1794), who discovered citrus as a potential therapy in 1753, and Lieut. James Cook's use of citrus on his Pacific expeditions prompted physicians to emphasize the value of citrus in the British Admiralty.[15] The Admiralty gave sailors daily lemon juice draughts, which they mixed with their rum ration. Nelson had a strategic edge over Napoleon's fleet at the Battle of Trafalgar because of the better health of the British navy. It is an antioxidant, however, and stimulates collagen synthesis, as well as increasing vasodilation and lowering the risk of a heart attack in individuals with excessive cholesterol and artery hardening. Although hesperidium has less vitamin C than many other fruits, citrus has a superior natural packing. Is this vitamin supply a reward for dispersion agents in the wild, or something else? It can only be synthesized by primates and guinea pigs.

9.3 SOME REPRESENTATIVE MEDICINAL PLANTS FROM THE RUTACEAE FAMILY

The rue family (Rutaceae) has about 2000 species in 160 genera, with the majority of them being woody shrubs and trees. The blooms are usually large and fragrant, and many species have fragrant leaves. Several of these plants, especially those in the *Citrus* genus, are significant food crops, while others are planted as garden ornamentals.

9.3.1 *CITRUS AURANTIUM*

C. aurantium L., often recognized as bitter orange, laranjeira-amarga, or laranjeira-cavalo, is a southeast Asian native plant. The Arabs were the ones who originally introduced it to Syria and Egypt, and then to Europe.[34] It was widely used in the form of tranquilizer, vasostimulant, sedative, orexigenic, digestive, antidote, and general tonic in the Mediterranean region throughout medieval times. The *C. aurantium* effects on the CNS, particularly its minor tranquilizer impact, have been the subject of research. The anxiolytic effect of this species has been demonstrated in several investigations on animals and humans. Anxiolytic impacts of *C. aurantium* peel essential oil (EO) were observed in rats following a single dose of the EO, which was given to them in the open arms of the plus maze.[7]

9.3.1.1 CLINICAL APPLICATION

Clinical experiments examining *C. aurantium's* anxiolytic impact have also shown positive findings. Preoperative patients were given distilled flowers of *C. aurantium* 2 h before the surgery, and the Spielberger STAI-state scale was used to measure anxiety. Patients in the experimental group had less anxiety before surgery than those in the comparison group.[1] Before the operation of collecting medullary material, patients with CML smelled *C. aurantium* EO. The results revealed that the patients who received this intervention had lower anxiety levels and stayed comfortable during the operation. Furthermore, compared to the anxiolytic used as a control, *C. aurantium* oil was equal in activity, even after a single treatment. This guarantees the effectiveness of oil in reducing anxiety in individuals undergoing hostile diagnostic processes.[20]

9.3.2 PHELLODENDRON AMURENSE

The Amur cork tree, also known as Huang Bai, is widely consumed in Chinese traditional medicine, and included in the 50 fundamental herbs; however, it must be taken with caution. The bark is a bitter medicine that has a powerful effect on the renal organ as well as being used to remove toxins.[33] The plant has recently been found to be effective in conjunctivitis and meningitis treatment. This herb must be taken with the guidance of a doctor and should not be used during pregnancy.[33]

There are a variety of health benefits associated with Huang Bai bark, including tonic, aphrodisiac, diuretic, antibacterial, bitter stomachic, antirheumatic, expectorant, skin, hypoglycemic, cholagogue, ophthalmic, febrifuge, and vasodilator. It's used to treat diarrhea, dysentery, and vaginal infections like *Trichomonas*, jaundice, acute UTI, abscesses, enteritis, night sweats, skin diseases, and boils.[26] It is commonly utilized with *Coptis chinensis* and *Scutellaria baicalensis* as "injection of three yellow herbs." It is administered intramuscularly for URI. In winter or spring, the bark of 10-year-old trees is collected and dried up for later purposes.[26] The fruits have a linctus effect.

9.3.2.1 CLINICAL APPLICATION

Abdominal discomfort, diarrhea, gastroenteritis, UTI, and other illnesses have all been treated using *Phellodendron amurense*. Berberine, obacunone, and obaculactone are the three main ingredients.[18] Obaculactone inhibits the alloantigen-specific production of T-helper cell 1 (Th1) cytokines, IFN, proinflammatory cytokines, TNF, IL-2, and IL-6 in mice receiving skin allografts, which is a unique immunomodulatory characteristic.[32] In terms of decreasing skin thickness and regulating cytokine production, the improvements in symptoms following therapy with *Phellodendron amurense* combination were more pronounced than those following treatment with HC. Finally, synergistic therapy with *Phellodendron amurense* may be a viable alternative therapeutic strategy for atopic dermatitis management.[18] *Epicoccum nigrum* QX501, an endophytic fungus isolated from the cambium of *Phellodendron amurense,* extracellularly produces Ag NPs. The capacity to manufacture Ag NPs as possible antibacterial agents using *E. nigrum* is extremely promising for the green, long-term production of nanometals, and it also expands the strategy's use.[21]

9.3.3 ZANTHOXYLUM

Zanthoxylum is a genus of around 200 fragrant trees and shrubs belonging to the Rutaceae family and is inherent to South America, North America, Asia, Australia, and Africa. Numerous species are grown for their beautiful wood or as ornamentals, and some are utilized in herbal medicine. The dried husks of several species, particularly *Zanthoxylum piperitum, Z. simulans*, and *Z. bungeanum*, are used to make Sichuan pepper, an Asian spice. Prickly ash is another name for the unrelated angelica tree, usually famous as the devil's walking stick (*Aralia spinosa*). For thousands of years, *Zanthoxylum* has been used as a hemostatic medication as well as to treat inflammation and pain, the medicinal part of which is the dried roots.

According to TCM, *Zanthoxylum* can relieve pain by promoting circulation, dispersing wind to dredge collaterals, promoting blood circulation to remove blood stasis, and removing toxicity to achieve detumescence; thus, it could be used to treat toothaches, stomachaches, traumatic injury, rheumatic arthralgia, and snake bites.[4] Inflammatory disorders, bacterial and viral infections, liver damage, cancers, gastric, and oral ulcers are among the other medicinal uses.[14]

9.3.3.1 CLINICAL APPLICATION

Amides, essential oils, lignans, flavonoids, and alkaloids are among the *Z. bungeanum* Maxim. components. The anesthetic effect is attributed to alkylamides, such as sanshools. It can decrease excitability and cause "fast pain analgesia" by targeting VGSCs on mechanosensory pain receptors. As a result, *Zanthoxylum* plants are known in Western folk medicine as "toothache trees." Alkylamides derived from *Zanthoxylum* are often used as odontologic agents, a toothache, buccal anesthetic, and sore throat painkiller.[4] HAS, an alkyl amide that is abundant in the pericarp of *Zanthoxylum*, is liable for the tingling/prickling and buzzing sensations linked to anesthetic characteristics. The prickling sensation induced by HAS (25–50 μg) was confined to the tongue and lasted 10–20 min when applied unswervingly to the human tongue. Minor surgical treatments can benefit from this anesthetic effect.[24]

9.3.4 AEGLE MARMELOS

Bael, or *Aegle marmelos* L., is an anti-inflammatory, antioxidant, and ulcer-fighting plant. In addition, research has shown that *A. marmelos* flower extract (AMF), an Indian traditional medicinal herb, was utilized to treat wounds.[2] Because of its numerous therapeutic qualities, *A. marmelos* Correa is frequently utilized in indigenous Indian medical systems. Despite its long history as an antidiarrheal, nothing is known about how it works in infectious diarrhea. Numerous researches have investigated the antidiarrheal activity of *A. marmelos* unripe fruit. *A. marmelos* is helpful in chronic diarrhea because of its high amounts of mucilage, which acts as a demulcent. *A. marmelos* has also been proven to be beneficial in animal models of irritable bowel syndrome and physiological diarrhea. Apart from antiprotozoal research, we are unaware of any further studies.

9.3.4.1 CLINICAL APPLICATION

Vast clinical and experimental research findings show that *A. marmelos* has antidiarrheal, antiseptic, antiviral, antimutagenic, anticancer, cytoprotective, antiparasitic, ulcer healing, antigenotoxic, diuretic, antineoplastic, and anti-inflammatory characteristics, allowing it to prevent the occurrence and treat a wide range of diseases. As a result, it is useful to evaluate its medicinal effects to provide an understanding of its position to both current and ancient scientists.[22]

9.3.5 *DICTAMNUS ALBUS*

The gas plant (*Dictamnus albus L.*) is a widespread plant found all over the world. *D. albus* grows wildly in Turkey, particularly in the northern Anatolian region. In Greece's flora, the genus *Dictamnus* (Rutaceae) has only one member, *Dictamnus albus* L. Flavonoids, quinoline alkaloids, psoralens (coumarins), dictagymnin, feniculin, and methylchavicol, with trans-anethol are the major constituents within essential oil. *D. albus* is a plant being used as an emmenagogue as well as an abortive agent. Its mutagenic and embryotoxic effects appear to be linked toward this abortive activity. This herb has been used in traditional medicine to treat jaundice as well as skin diseases such as psoriasis. *D. albus* L. and *D. hispanicus* Webb ex Willk. are members of the genus *Dictamnus*. During 1000 and 1700 C.E., a variety of complicated concoctions including *Dictamnus* were employed to treat 35 various diseases. The common *D. albus* is a considerably better plausible prototype *Dictamnus* than the Cretan unique *Origanum dictamnus* for biogeography reasons. Both, nevertheless, are essential components of the *Dictamnus* complex. There is an indication that *D. albus* and *D. hispanicus* have a long and consistent history of being used to cure 47 different illnesses.[16]

9.3.5.1 CLINICAL APPLICATION

The psoralen concentration of the gas plant causes photodermatitis. Despite significant research into the medical applications and phototoxicities of psoralens, the method of their action within the epidermis is yet to be determined. The production of reactive oxygen species and subsequent

DNA damage, which leads to cell death or mutagenesis, is one proposed explanation for their phototoxic effects.[31] Open patch skin tests in guinea pigs verified prior reports of photodermatitis induced by the gas plant in human subjects in the current investigation. *Dictamnus albus* aqueous extracts have effects on the central nervous system. In "tail-clip" studies, extracts of *D. albus* were shown to be ineffective, implying that they had neither algesic nor analgesic effects at the dosage used.[28] They did, however, produce a substantial reduction in the swimming time of mice, indicating that they had a depressive effect on the CNS. This was verified in our general behavior screening studies, which revealed a clear calming effect and a decrease in spontaneous behaviors in mice after exposure to plant extracts.[3]

9.3.6 PTELEA TRIFOLIATA

Ptelea trifoliata L. (Rutaceae), often known as "hoptree" or "wafer ash," is a dioecious big shrub or small tree that grows up to 6 m tall. *P. trifoliata* is found throughout eastern North America, from Wisconsin to New York, south to Florida and Mexico, and west to Texas and Kansas, including isolated populaces as far west as Colorado. The huge swallowtail butterfly, *Papilio cresphontes* Cramer, feeds on the leaves of *P. trifoliata*. It has been used as an antiperiodic, anthelmintic, stomachic, antimicrobial, and tonic in herbal medicine. The fruits were previously employed as a hops replacement in beer production.[30]

9.3.7 FORTUNELLA

The kumquat tree is a tiny tropical fruit-bearing tree. It originates in China, where it has been grown since the 12th century. Its name comes from the Cantonese Chinese word "kam kwat," which means "golden orange," since the fruit resembles a little orange. Kumquats were formerly classed as a member of the *Citrus* genus, but after acclimating to European conditions, they were given their own genus, *Fortunella*.[6] The facts and data gathered show that this plant has tremendous chemical, biological, and nutritional potential. The antioxidant, antibacterial, anticancer, and anti-inflammatory properties of kumquat and its components have been studied, but little attention has been paid to the discovery of bioactive chemicals in the fruit.[31] Phenolic chemicals, particularly flavonoids, are the most well-studied chemical components of kumquat in the literature. The antioxidant activity of kumquat essential oils is influenced by its components; the most important chemicals thought to have such an impact are as follows: α-terpinene, nootkatone, citronellal, citral, geraniol, γ-terpinene, and terpinolene are all derivatives of terpinene.[13]

9.3.7.1 CLINICAL APPLICATION

Bioactivities such as antioxidant, anticancer, antibacterial, and anti-inflammatory are due to the presence of numerous active chemicals. In addition to its health benefits, kumquat has the potential to be used as a raw material in culinary and cosmetic applications. It may be used in the culinary sector to make sweets, desserts, jams, jellies, and most importantly, liqueurs. The whole fruit, including the peel, functions as a slimming agent, regulating metabolism and cholesterol levels, vitaminizing, and exhibiting antibacterial and anti-free radical properties.[32] Many applications are determined by these actions and characteristics. Kumquat fruits are a high-potential natural raw material that may be used in functional foods, para pharmaceutical goods, and cosmetics.[19]

9.3.8 CHLOROXYLON SWIETENIA

Chloroxylon swietenia DC, a Rutaceae family plant, is a medium-sized deciduous tree with a spreading crown that grows to a height of 9–15 m and a girth of 1.0–1.2 m. Ceylon Satinwood or East Indian Satinwood is the name given to a tree that is indigenous to Sri Lanka and India. *C. swietenia* is a plant that has been utilized in traditional treatments for a variety of ailments. In folklore medicine, *C. swietenia* has been employed. Wounds, cuts, burns, and skin disorders are treated with leaf paste by the Malasar tribes of Coimbatore (Tamil Nadu, South India).[10] The leaf paste is applied by Chenchus in Nallamalais (Andhra Pradesh, South India) to cure worm-infested animal wounds, fungal skin infections, and rheumatism. Snakebites, common colds, coughs, ocular infection, cataracts, astringent, itches, headaches, impotence, and other conditions are all treated with various components of the plant.[11]

9.3.8.1 CLINICAL APPLICATION

C. swietenia has been employed in folklore medicine, and its therapeutic use has been widespread throughout history. Some of its traditional uses have been scientifically validated, but others have yet to be done. The preservation of this medicinally useful tree is critical.[5]

9.3.9 CITRUS RETICULATA

Citrus reticulata "Chachi" is a tangerine species (i.e., taxon) in the *Citrus reticulata* genus. The Chinese Pharmacopoeia lists it's dehydrated and matured peel Pericarpium *Citri reticultae* (Guang Chenpi) as suitable for medicinal usage. In China, *C. reticulata* "Chachi" is consumed as a spice and a tea component. Pericarpium *Citri reticultae* possesses antioxidant, antimutagenic, anticancer, anti-inflammatory, and antitumor actions, according to pharmacological studies.[8] The function of antiatherosclerosis agent is to decrease phlegm in the lungs. The major chemical components of Pericarpium *Citri reticultae* are known to be flavonoids such as didymin, naringin, tangeretin, nobiletin, and hesperidin. Flavonoids have recently got a lot of attention, and some studies have shown that they might play a key role in anticancer activity.[29]

9.4 SUMMARY

Rutaceae family is well-known and widely used for a variety of purposes by people since time memorial. Because of their medicinal potential, numerous researches have been conducted to demonstrate the pharmacological effects of the aforementioned species. However, research into alternative pharmacological effects, particularly of isolated components of the species, has to be progressed.

ACKNOWLEDGMENTS

The authors are very thankful to all the authors whose work has been cited in this chapter.

KEYWORDS

- **citrus**
- **Rutaceae**
- **clinical trial**
- **pharmacological activity**
- **contraception**
- **scurvy**

REFERENCES

1. Akhlaghi, M.; Shabanian, G.; Rafieian-Kopaei, M.; Parvin, N.; Saadat, M.; Akhlaghi, M. *Citrus aurantium* Blossom and Preoperative Anxiety. *Revista Brasileira de Anestesiologia* 2011, *61* (6), 707–712.
2. Azmi, L.; Shukla, I.; Goutam, A.; Rao, C. V.; Jawaid, T.; Kamal, M. *In Vitro* Wound Healing Activity of 1-hydroxy-5, 7-dimethoxy-2-naphthalene-carboxaldehyde (HDNC) and Other Isolates of *Aegle marmelos* L.: Enhances Keratinocytes Motility via Wnt/$β$-Catenin and RAS-ERK Pathways. *Saudi Pharm. J.* 2019, *27* (4), 532–539.
3. Beis, R.; Öztürk, Y.; Flores, A.; Gürer, F. Pharmacological and Toxicological Effects of Gas Plant (*Dictamnus albus* L.). *Turkish J. Pharm. Sci.* 2005, *2* (3), 111–124.
4. Bhattacharya, S.; Zaman, M. K.; Haldar, P. K. Antibacterial Activity of Stem Bark and Root of Indian *Zanthoxylum nitidum*. *Asian J. Pharm. Clin. Res.* 2009, *2* (1), 30–34.
5. Charanraj, N.; Venkateswararao, P.; Vasudha, B.; Narender, B. Phytopharmacology of *Chloroxylon swietenia*: A Review. *J. Drug Deliv. Ther.* 2019, *9* (1), 273–278.
6. Chen, M. H.; Yang, K. M.; Huang, T. C.; Wu, M. L. Traditional Small-Size Citrus from Taiwan: Essential Oils, Bioactive Compounds and Antioxidant Capacity. *Medicines* 2017, *4* (2), 28.
7. de Moraes Pultrini, A.; Galindo, L. A.; Costa, M. Effects of the Essential Oil from *Citrus aurantium* L. in Experimental Anxiety Models in Mice. *Life Sci.* 2006, *78* (15), 1720–1725.
8. Imbesi, A.; De Pasquale, A. Citrus Species and their Essential Oils in Traditional Medicine. In *Citrus: The Genus Citrus*; Dugo, G., Di Giacomo, A., Eds.; CRC Press, 2002; pp 591–615.
9. Jayaprasad, B.; Sharavanan, P. S.; Sivaraj, R. Antidiabetic Effect of *Chloroxylon swietenia* Bark Extracts on Streptozotocin Induced Diabetic Rats. *Beni-Suef Univ. J. Basic Appl. Sci.* 2016, *5* (1), 61–69.

10. Kiran, S. R.; Devi, P. S. Evaluation of Mosquitocidal Activity of Essential Oil and Sesquiterpenes from Leaves of *Chloroxylon swietenia* DC. *Parasitol. Res.* 2007, *101* (2), 413–418.
11. Lou, S. N.; Ho, C. T. Phenolic Compounds and Biological Activities of Small-Size Citrus: Kumquat and Calamondin. *J. Food Drug Anal.* 2017, *25* (1), 162–175.
12. Lu, Q.; Ma, R.; Yang, Y.; Mo, Z.; Pu, X.; Li, C. *Zanthoxylum nitidum* (Roxb.) DC: Traditional Uses, Phytochemistry, Pharmacological Activities and Toxicology. *J. Ethnopharmacol.* 2020, *260*, 112946.
13. Mabberley, D. J. Citrus (Rutaceae): A Review of Recent Advances in Etymology, Systematics and Medical Applications. *Blumea-Biodivers. Evolut. Biogeog. Plants* 2004, *49* (2–3), 481–498.
14. Martinez-Frances, V.; Rivera, D.; Heinrich, M.; Obón, C.; Rios, S. An Ethnopharmacological and Historical Analysis of "*Dictamnus*", A European Traditional Herbal Medicine. *J. Ethnopharmacol.* 2015, *175*, 390–406.
15. Park, S.; Kim, D. S.; Kang, S.; Shin, B. K. Synergistic Topical Application of Salt-Processed *Phellodendron amurense* and *Sanguisorba officinalis* Linne Alleviates Atopic Dermatitis Symptoms by Reducing Levels of Immunoglobulin E and Pro-inflammatory Cytokines in NC/Nga Mice. *Mol. Med. Rep.* 2015, *12* (5), 7657–7664.
16. Pawełczyk, A.; Żwawiak, J.; Zaprutko, L. Kumquat Fruits as an Important Source of Food Ingredients and Utility Compounds. *Food Rev. Int.* 2021, 1–21.
17. Pimenta, F. C. F.; Alves, M. F.; Pimenta, M. B. F.; Melo, S. A. L.; Almeida, A. A. F. de; Leite, J. R.; ... Almeida, R. N. de. Anxiolytic Effect of *Citrus aurantium* L. on Patients with Chronic Myeloid Leukemia. *Phytother. Res.* 2016, *30* (4), 613–617.
18. Qian, Y.; Yu, H.; He, D.; Yang, H.; Wang, W.; Wan, X.; Wang, L. Biosynthesis of Silver Nanoparticles by the Endophytic Fungus *Epicoccum nigrum* and their Activity Against Pathogenic Fungi. *Bioproc. Biosyst. Eng.* 2013, *36* (11), 1613–1619.
19. Rahman, S.; Parvin, R. Therapeutic Potential of *Aegle marmelos* (L.)-An Overview. *Asian Pac. J. Trop. Dis.* 2014, *4* (1), 71–77.
20. Raizman, N.; et al. Scurvy: How a Surgeon, a Mariner, and a Gentleman Solved the Greatest Medical Mystery of the Age of Sail. *J. Clin. Investig.* 2004, *114* (12), 1690.
21. Rong, R.; Cui, M. Y.; Zhang, Q. L.; Zhang, M. Y.; Yu, Y. M.; Zhou, X. Y.; ... Zhao, Y. L. Anesthetic Constituents of *Zanthoxylum bungeanum* Maxim.: A Pharmacokinetic Study. *J. Sep. Sci.* 2016, *39* (14), 2728–2735.
22. Sheat, W. G. In *Propagation of Trees, Shrubs and Conifers*; Macmillan and Co.: London, 1948; vol. XII, p 479.
23. Short, R. V.; McCoombe, S. G.; Maslin, C. L. V.; Crowe, S. M. In *Lemon and Lime Juice as Potent Natural Microbicides*, AIDS 2004: Proceedings of the XV International AIDS Conference 2004, 2004.
24. Sun, Y.; Wang, J.; Gu, S.; Liu, Z.; Zhang, Y.; Zhang, X. Simultaneous Determination of Flavonoids in Different Parts of *Citrus reticulata* 'Chachi'Fruit by High Performance Liquid Chromatography—Photodiode Array Detection. *Molecules* 2010, *15* (8), 5378–5388.
25. Takaku, S.; Setzer, W. N. Chemical Composition of the Leaf Essential Oil of *Ptelea trifoliata*. *J. Essent. Oil Bear. Plants* 2007, *10* (2), 104–108.

26. Woo, W. S.; Lee, E. B.; Han, B. H. Biological Evaluation of Korean Medicinal Plants (III). *Arch. Pharm. Res.* 1979, *2* (2), 127–131.
27. Xian, Y. F.; Mao, Q. Q.; Ip, S. P.; Lin, Z. X.; Che, C. T. Comparison on the Anti-inflammatory Effect of Cortex Phellodendri Chinensis and Cortex *Phellodendri amurensis* in 12-O-tetradecanoyl-phorbol-13-acetate-Induced Ear Edema in Mice. *J. Ethnopharmacol.* 2011, *137* (3), 1425–1430.
28. Yeung, H. C. In *Handbook of Chinese Herbs and Formulas Institute of Chinese Medicine*, 2nd Ed.; Redwing Book Co., 1985; p 431.
29. Zhi-Yao, W.; Wen-Jun, H. E.; Wen-Bing, Z.; Guang-Zhi, Z.; Zhi-Qi, Y. I. N.; Shou-Xun, Z.; Ning-Hua, T. A. N. Two New Phenylpropanoids from *Micromelum integerrimum*. *Chinese J. Nat. Med.* 2014, *12* (8), 619–622.
30. Zou, Z.; Xi, W.; Hu, Y.; Nie, C.; Zhou, Z. Antioxidant Activity of *Citrus* Fruits. *Food Chem.* 2016, *196*, 885–896.

CHAPTER 10

Toxicological Profile of the Rutaceae Family

CHENMALA KARTHIKA[1], SHROUQ H. SWEILAM[2,3], MUDDASER SHAH[4], and MD. HABIBUR RAHMAN[5,6]

[1]*Department of Pharmaceutics, JSS College of Pharmacy, JSS Academy of Higher Education and Research, Ooty, Nilgiris, Tamil Nadu, India*

[2]*Department of Pharmacognosy, Faculty of Pharmacy, Prince Sattam Bin Abdulaziz University, Al-Kharj, Saudi Arabia*

[3]*Department of Pharmacognosy, Faculty of Pharmacy, Egyptian Russian University, Badr City, Cairo, Egypt*

[4]*Department of Botany, Abdul Wali Khan University Mardan, Mardan, Pakistan*

[5]*Department of Global Medical Science, Yonsei University Wonju College of Medicine, Yonsei University, Wonju, South Korea*

[6]*Department of Pharmacy, Southeast University, Banani, Dhaka, Bangladesh*

ABSTRACT

The toxicology profile of the Rutaceae family is discussed in this chapter. The Rutaceae family's species have been shown to have medical benefits and have been widely used since ancient times. Plants of this type can also be found in Asia. Despite the fact that plants in the Rutaceae family have

Phytochemical and Pharmacological Investigation of the Family Rutaceae.
Abdur Rauf, Hafiz Ansar Rasul Suleria, & Syed Muhammad Mukarram Shah (Eds.)

been shown to have a variety of therapeutic effects, they also have some toxic effects. Toxic effects are typically caused by compatibility issues or the administration of a high dose. Despite the fact that the plant's various therapeutic benefits have been demonstrated and validated since antiquity, its toxicity and associated issues have received little attention. This chapter puts forward some of the proven toxic effects of plants belonging to the Rutaceae family. The studies are done mainly on the experimental and not entered into the clinical phase.

10.1 INTRODUCTION

Rutaceae is a flowering plant family comprised of approximately 160 genera and species. Another name for it is the citrus family. This category includes trees, lianas, as well as, on rare occasions, herbs.[1] The Rutaceae family is divided into four subgroups. Some taxa have thorns on their stems. They are simple, trifoliolate, or pinnate, with pellucid or punctate, on rare occasions, pinnatifid as well as exotype-late. On the inflorescence, there are usually a few solitary flowers.[2] In tropical areas, the Rutaceae family has a large number of members. Among the most economically important fruits are citrus species (orange, grapefruit, lemon, and lime), as well as *Ruta graveolens* (rue), timber trees, medicinal plants, and a variety of ornamental cultivars.[3–5]

There are 250 species found in the Rutaceae family, with 45 species and 13 variants found in China. *Zanthoxylum bungeanum* is a species of *Zanthoxylum* found in China, India, Japan, and Korea. The fruit of *Zanthoxylum bungeanum* has been used in Chinese cuisine as well as medicine for centuries.[6–8] Two of the many variants that have been cultivated are Yuexi gong jiao and Hanyuan Huajiao. Because of their bright red pericarps, *Z. bungeanum* cultivars are known as "Honghua Jiao" in Chinese.[9,10] The plant list database includes evergreen shrubs and trees from tropical low and evergreen forests, as well as some submontane forests (www.theplantlist.org). "*Vepris* Comm. Ex. A. Juss" species can be found all over the world, from the dense forest of Guinea on the edge of the Sahara Desert to the dense forest of South Africa.[11] *Vepris lanceolata*, a popular conventional medicinal plant in Tanzania, Mauritius, Reunion, and South Africa, has been reported to help with wounds, stomachaches, fevers, sores, influenza, pains, sprains, rheumatic pains, malaria, dysentery, catarrh, headaches, as well as other ailments.[12,13]

Agathosma (Rutaceae) is a plant genus in South Africa with approximately 150 species, possessing woody stems as well as small, flat, convex to ericoid leaves. They've been dubbed "Buchu" ever since. It is a well-known herb in South Africa as well as throughout the world (*Agathosma betulina* and *Agathosma crenulata*). The medicinal plant "Buchu" was introduced to Europeans by the indigenous people of the Cape. "Buchu" was widely accepted as a medicine in Europe and America, and used for many years in both countries.[14,15] The plants in the Rutaceae family that have been tested for toxicity are listed below.

10.2 TOXICOLOGY PROFILE

10.2.1 ZANTHOXYLUM BUNGEANUM MAXIM. (RUTACEAE)

Z. bungeanum is currently prescribed in the People's Republic of China's Pharmacopoeia (Ch. P.) for indigestion, dyspepsia, vomiting, diarrhea, ascariasis, dermatitis, as well as other disorders.[16] More than 140 chemical compounds have been discovered and described as a result of the growing corpus of studies on *Z. bungeanum. Z. bungeanum* has antioxidant, antifungal, analgesic, antibacterial, anti-inflammatory, and antitumor characteristics, as well as effects on proper circulation and digestion.[17,18] Since ancient times, *Z. bungeanum* has been used in conventional Chinese medicine.[19] There have been a few studies on *Z. bungeanum's* toxicity, with previous research focusing on its extracts (Table 10.1). In a 1995 investigation, Tong et al. discovered the median lethal concentration (LD_{50}) of WEZB in mice (i.g., equivalent to crude herbs mass). According to a 2010 study, the toxicity of *Z. bungeanum* varies depending on genetic features, growing circumstances, and medicinal components. Zhao et al. discovered in 2003 that rats fed WEZB at doses of 0.5, 1, 2, and 4 g/kg (equivalent to comparable amounts of crude herbs) suffered minor liver injury, including ballooning degeneration, cytoplasm rarefaction, and patchy necrosis.[20–22] J774.1 cells were given WEZB at concentrations of 100, 200, and 400 g/mL. According to the findings, WEZB had no toxicity at these doses. Toxicology has also been studied in recent years. Intragastric administration (IGA), intraperitoneal (IPI) injections, intramuscular (IMI) injections, and hypodermic injections (HID) are the four methods for administering the EOZB's LD50 value (i.h.). Mice were given lethal doses of EOZB and they died as a result. EOZB is nontoxic to HaCaT and CCC-ESF-1 cells, with IC_{50} values of 2.435 and 3.649 mg/mL, respectively.[23–25]

Z. bungeanum was discovered to have a low toxicity potential when used as a flavoring or in conventional Chinese medicine. Overdosage or misuse of *Z. bungeanum* was the most common cause of adverse reactions. Overdosing of this plant can lead to serious side effects, if not death. If you are pregnant or breastfeeding, or have a Yin deficiency, you should avoid taking *Z. bungeanum*.[26]

TABLE 10.1 Studies on the Toxicity of *Z. Bungeanum*.

Extracts or compound	In vitro or in vivo	Dose or minimal toxic concentration	Toxic effect
Water extracts of *Z. bungeanum*	Mice	LD_{50} = 45 g/kg (i.g., equivalent to crude herbs mass)	Mortality
Water extracts of *Z. bungeanum*	Mice	LD_{50} value is 51.14 g/kg (i.g., equivalent to crude herbs mass)	Mortality
Water extracts of *Z. bungeanum*	Mice	0.5, 1.0, 2.0 and 4.0 g/kg (i.g.)	Ballooning and degeneration
Essential oils of *Z. bungeanum*	Mice	LD_{50} = 2.27, 2.03, 4.64, and 5.32 g/kg of i.g., i.p., i.m., and i.h., respectively	Mortality
Essential oils of *Z. bungeanum*	HaCaT cells and CCC-ESF-1 cells	IC_{50} = 2.435 mg/mL and 3.649 mg/mL, respectively	Inducing cell viability
Water extracts of *Z. bungeanum*	J774.1 cells	100, 200, and 400 g/mL (for 18 h)	Nontoxic effect

Z. bungeanum has been used in conventional Chinese medicine as a food component for a long time. Several characteristics of this plant have improved dramatically in the past 10 years. Before clinical research can be regarded as successful, a few barriers must be addressed. This plant's pharmacological effect raises a variety of concerns.[26,27] As a result, additional research is needed to completely understand the structure–function relationship of bioactive chemicals as well as their methods of action. Another issue is the paucity of pharmacokinetics as well as clinical investigations on *Z. bungeanum* and analyses of cellular and molecular toxicity. Thus, the future *Z. bungeanum* research should concentrate on the pharmacokinetics of compounds other than alkylamides, as well as molecular and cellular toxicity investigations to explore the potential adverse effects in clinical trials.[28–33] HAS appears to be the principal active molecule in both in vivo and in vitro testing. HAS, on the other hand, is unstable in normal storage conditions, as well as due to its triene system, it may be sensitive to oxygen. By changing the structure of HAS and other sanshools, more stable derivatives can be created. Conventional *Z. bungeanum* applications do not include prescriptions, and processed *Z. bungeanum* products are becoming increasingly scarce. As a result, new prescriptions and products are required to meet the clinical needs. Research on *Z. bungeanum* seeds, stems, and roots has lagged behind pericarp research; in order to fully exploit this plant's medicinal potential, it is essential to explore the chemical composites as well as pharmacological properties of every part of *Z. bungeanum*. *Z. bungeanum*'s relatives are difficult to distinguish due to their rich biodiversity and morphological similarities to *Zanthoxylum*. This includes the creation of a potential plant quality control strategy as well as the establishment of an international quality evaluation system that all parties can use. Finally, because it is widely distributed and grown in many different parts of the world, it is critical to improve the quality and efficiency of the picking process.[34,35]

10.3 GENUS VEPRIS (RUTACEAE)

An ethanolic extract from the root of *V. glomerata* from Tanzania yielded a fatal concentration of 51.05 g/mL when tested on the brine shrimp *A. salina* (LC_{50}). A total of 10 mice were separated into 10 groups and fasted for 24 h before the trial in an acute toxicity test.[36–38] All groups were given solvent as well as 2400 mg of extract/kg of body weight the following day. For 24 h, the mice were kept under observation. Following that, a mouse *A. salina* experiment indicated no negative effects. Citral Extrasynthese S.A. (Genay, France) sells a nontoxic to normal Chang liver cells product derived from *V. macrophylla* leaf oil. Toxicological tests were carried out on the root extract (apparently *V. glomerata*'s root) collected in Tanzania the same year.[39–42] The extract was unaffected by 2400 mg/kg body weight. A fraction of a dichloromethane/methanolic extract from the roots of *V. uguensis* containing the flinders amine paralyzed the housefly *M. domestica* in 3.98 min. When combined with the insecticide flinders amine, positive aerosols developed by the Kenya Bureau of Statistics were discovered to have insecticidal activity synergistic with each other,[39] killing at a rate of 95% after 24 h. During ethnobotanical research on plants used in rural Uganda that same year, an extract of *V. nobilis* bark was tested against the MRC-5 cell line, which inhibited it by 22.0% (no positive control was provided).

In *Côte d'Ivoire*, the stem bark of *V. verdoorniana* is used as a fish poison, indicating that it is highly toxic to other multicellular organisms. Young as well as middle-aged mouse hookworm *H. bakeri* larvae were fed a *V. louisii* stem bark extract in petri dishes. The extract's LC_{50} was 2398.83 g/mL, while albendazole, a strong anthelmintic, had LC_{50} of 2278.98 and 2772.39 g/mL, respectively. Furthermore, the new compounds 6,7-methylenedioxy-5-hydroxy-8-methoxy-dictamnine and 5-hydroxy-4′-methoxy-7-O-[-L-rhamnopyranosyl (1‴5″)-dapiofuranosyl] were found to be toxic in an MTT toxicity assay. Flavanoside, evoxanthine, roeharmine, 3-hydroxy-olean-12-ene, didymin and asscoparone had IC50 values of 0.67, 10.5, 10.4, 10.9, 7.6, and 12.6 M, respectively, on the MRC-5 cell line, compared to vinblasine's IC50 value of 2.8 M.

T. castaneum and the red flour weevil were tested for contact toxicity using the essential oil of *V. heterophylla* leaves. According to leaf essential oil samples collected in Mokolo and Meri, Cameroon, killing half the population in a single day would require 49.44 and 61.2 L of leaf essential oil, respectively. The high level of toxicity could be attributed to the highly toxic oil ingredients in the Mokolo sample.

The toxicity of *V. heterophylla* hexane leaf extracts against the red flour weevil *T. castaneum* adults was investigated as part of the study. After 72 h of being exposed to 0.75 g/mL in petri dishes, 99.5% of the weevils died. After ingesting powdered *V. heterophylla* leaves, more than half of the *T. castaneum* larvae died. The cytotoxicity of maculine B, kokusaginine B, and veprisazole was investigated using the AML12 cell line. The IC_{50} values for the new compounds were 96.54, >154.4, and 62.5 M, respectively, compared to 0.48 M for doxorubicin.[43–45] Table 10.2 shows the compound's toxicity and its mode of action.

10.4 AGATHOSMA (RUTACEAE) SPECIES

The MTT test was performed to determine the toxicity found in the *Agathosma* species. The most toxic plants were *Agathosma lanata* (IC_{50} of 26.17 ± 9.58 g/mL) and *Agathosma ovata* (IC_{50} of 25.20 ± 6.30 g/mL). The IC_{50} values for *Agathosma roodebergensis*, *Agathosma collinus*, hirsute *Agathosma* strains, as well as *Agathosma zwartbergense* ranged from 38.05 to 54.68 g/mL.[46–48] *Agathosma crenulata*, *Agathosma betulina*, *Agathosma capensis* (Gamka), and *Agathosma namaquensis* were all found to be safe at dosages of up to 100 g/mL. Sample levels were found to be hazardous in a dose-dependent

TABLE 10.2 Toxicity of the Compounds Extracted from Genus *Vepris*.

Sl. No.	Year	Author as well as year	Compound	Method	Action
1	2010	Mwangi et al.	Skimmianine	Alamar blue	IC_{50} = 38.6 μg/mL against L6 cell line
2	2016	Nouga et al.	γ-fagarine	MTT	IC_{50} against MRC-5 cell line = 10.4 ± 2.7 M Vinblastine has an IC_{50} of 2.8 ± 0.4 M.
3	2019	Nganou et al.	Maculine B	Alamar blue	IC_{50} against AML12 cell line = 96.54 ± 3.09 M Doxorubicin has an IC50 of 0.48 ± 0.01 M.
4	2019	Nganou et al.	Kokusaginine B	Alamar blue	IC_{50}>154.44 M against AML12, doxorubicin IC50 = 0.48 M
5	2016	Nouga et al.	Evoxanthine	MTT	Vinblastine has an IC_{50} of 10.9 ± 1.2 M against the MRC-5 cell line and an IC50 of 2.8 ± 0.4 M.
6	2010	Mwangi et al.	Arborinine	Alamar blue	IC_{50} against L6 cell line = 12.2 g/mL
7	2010	Mwangi et al.	Normelicopicine	Alamar blue	The IC_{50} for the L6 cell line is greater than 90 g/mL.
8	2016	Nouga et al.	Roeharmine	MTT	Vinblastine has an IC_{50} of 7.6 ± 1.2 M against the MRC-5 cell line as well as an IC_{50} of 2.8 ± 0.4 M.
9	2019	Nganou et al.	Veprisazole	Alamar blue	IC_{50} against AML12 cell line = 62.44 ± 13.76 M Doxorubicin has an IC_{50} of 0.48 ± 0.01 M.
10	2016	Nouga et al.	3β-hydroxy-olean-12-ene	MTT	IC_{50} against MRC-5 cell line = 12.6 ± 2.4 M Vinblastine has an IC_{50} of 2.8 ± 0.4 M.
11	2010	Mwangi et al.	β-sitosterol	Alamar blue	IC_{50}>90 μg/mL against L6 cell line
12	2016	Nouga et al.	5-hydroxy-4′-methoxy-7-O-[α-Lrhamnopyranosyl(1‴→5″)-β-dapiofuranosyl]-flavanoside	MTT	Vinblastine has an IC50 of 10.5 ± 2.5 M against the MRC-5 cell line as well as an IC_{50} of 2.8 ± 0.4 M.

TABLE 10.2 *(Continued)*

Sl. No.	Year	Author as well as year	Compound	Method	Action
13	2016	Nouga et al.	Didymin	MTT	Vinblastine has an IC_{50} of 15.9 ± 3.2 M against the MRC-5 cell line as well as an IC50 of 2.8 ± 0.4 M.
14	2016	Nouga et al.	Scoparone	MTT	IC_{50} against MRC-5 cell line = 12.3 M Vinblastine has an IC_{50} of 2.8 M.
15	2009	Chaouki et al.	Geranial + Neral	MTT	IC_{50} against normal Chang liver cells > 150 M Doxorubicin was used as a control; however, the value was not published.

way. Simultaneous toxicology studies were conducted on all *Aridia* species, including *Arida*, *Collina*, *Horstus*, *Roodebergensis*, *Stitata*, and *Zartbergense*. Alkaloids were discovered in the aerial parts of *Agathosma* species by Campbell et al. (1987, 1990). Halfordamine as well as skimmianine were discovered in five species, including *A. capensis*. Quinoline alkaloids have been found in *A. barosmaefolia* as well as other plants.[49–52] Quinoline alkaloids have been found in *A. sativa*.[52] Plants like *Barosmaefolia* and others when tested against the ovarian cancer cell line A2780, seven alkaloids, including skimmianine, were found to have weak antitumor properties. According to Cheng et al., skimmianine decreased 5-hydroxytryptamine-induced vasopressin responses in rats, with greater dosages producing a nonspecific loss in cardiovascular function. It also reduced spontaneous motor activity, exploratory behavior, cataleptic reflexes, and long-term isolation-induced fighting in rats, according to another research. South Africans still employ "Buchu" as part of their healing culture, which traces back to the San and Khoi people of the Cape. It has been used to treat kidney infections, colds, and flu, as well as urinary tract infections and as a cough cure. Scientists may now securely propose *Agathosma* species to treat a wide range of diseases. These species have been found to contain antioxidant chemicals, which may (in part) contribute to their health benefits.[52–55]

10.5 CITRUS SPECIES—P-SYNEPHRINE

P-synephrine, commonly known as oxedrine, is an amine that is used all over the world to treat hypotension, and used as an ocular decongestant. The main active element is found in unripe *Citrus aurantium* L. fruits.[56] Ephedra alkaloids have gained appeal as an alternative to endogenous neurotransmitters for the treatment of obesity since the FDA prohibited them from dietary supplements in April 2004 due to their link to major health concerns (epinephrine as well as norepinephrine). Citrus species have undergone in vivo research.[57–59] In eight groups, mice were given water or 300, 500, 1000, 2500, 3500, or 5000 mg/kg of *C. aurantium* unripe fruits extract (2.5% p-synephrine) dissolved in 150, 300, 450, 600, 800, 1000, or 2000 mg/kg of p-synephrine.[60]

For all doses, the concentration was 10 mL/kg. Mice's respiratory, digestive, and neurological systems were thoroughly investigated. Every 24 h, fatalities as well as weights were recorded for 14 days.[61] A microscope was used to examine the hearts, livers, kidneys, and adrenal glass well ass of deceased animals. Necropsies were performed on the mice who survived the experiment until the 14th day. In the acute toxicological experiments, we employed eight male rats to reduce biological variability in 6–7 dose levels, assuring consistency in the reports seen with the drugs tested and allowing us to compute the LD50 threshold. The dosages of *C. aurantium* in the extract were estimated using a p-synephrine concentrator.[62,63]

After eating 1000–5000 mg/kg of *C. aurantium*, mice had a 2-h decrease in locomotor activity. Locomotor activity, piloerection, and exophthalmia were all reduced in p-synephrine-treated rats. Exophthalmia and piloerection occurred 15 min after receiving 300–2000 mg/kg p-synephrine and continued for 2 h. A 1-h effect of 300–2000 mg/kg p-synephrine on locomotor activity was observed. All doses studied resulted in salivation within 15 min, which lasted 30 min, as well as all amounts resulted in 3–4 h of gasping. In the control group, there were no anomalies.[64] There was no fatality because the LD_{50} limit for any of the doses tested could not be calculated. Both groups acquired the same amount of weight over the 14-day period. The excised organs exhibited no symptoms of injury during the necropsy.[65]

10.6 TOXICITY OF ESSENTIAL OILS OF FAMILY RUTACEAE: A STUDY IN PAKISTAN

Some Rutaceae species have become naturalized in Pakistan, where they are presently produced as well as hybridized for culinary, medicinal, and aesthetic purposes. The majority of the plant's leaves have fragrant glands, while some have thorns.[66] Due to the availability of essential oils with varied activities in members of the Rutaceae family, we undertook a complete assessment of the insecticidal and repellent properties claimed for essential oils in members of the Rutaceae family for the first time in Pakistan. The essential oils of each plant repulsed *Lasiusniger* (black ant). Each essential oil's repellent effects were found to differ slightly from one another, but significantly from the repellent effects of other concentrations as well as controls.[67,68] *M. paniculata* cv. Desi and *M. koenigii* had the highest insecticidal activity (6.58 L), followed by *M. paniculata* cv. China (8.41 L) and *Skimmia laureola* (8.41 L) (10.15 L).[69] All essential oils tested had a significant toxicity as well as repellent effect on adult garden ants (*Laciusniger*). This is due to the fact that this activity confirms previous findings. *M. koenigii* and *M. paniculata* cv. Desi essential oils had higher insect repellent properties than *M. paniculata* cv. China and

S. laureola essential oils.[70–73] In terms of phytochemical makeup, the essential oils primarily contained monoterpene and sesquiterpene. Monoterpenes were abundant in *M. koenigii*'s essential oil, but sesquiterpenes were abundant in *M. paniculata* cv. China and *M. paniculata* cv. Desi's essential oils. Monoterpenes have been identified as having acridity.[74] The toxic and repellent properties of essential oils are due to one or more of the compounds or chemical groups listed above. Nobody knows how terpenoids and essential oils fight bacteria, but lipophilic molecules may be involved. If essential oils have varying toxicity and repellence attributes, it could be due to the changes in the insect action targets of the compounds, as well as qualitative or quantitative variations.[75–80]

10.7 SUMMARY

Because plants in the Rutaceae family have long been used to cure a range of ailments, researchers are conducting comprehensive laboratory experiments. Surprisingly, the pharmacological properties of the Rutaceae family have been recognized, but there have been minimal toxicity studies. Plant extracts from the Rutaceae family have been tested for pharmacological effects in vitro as well as in vivo, but there is little information on their toxicological profile. The toxicity of the extracts from the Rutaceae family was not conducted further with the clinical trials, and also the toxicity is mainly dose dependent. More research needs to be focused on evaluating the toxicology profile of these plants.[80–85]

KEYWORDS

- **diseases**
- **dose**
- **extract**
- **pharmacology profile**
- **Rutaceae**
- **species**
- **toxicology**

REFERENCES

1. Kubitzki, K.; Kallunki, J. A.; Duretto, M.; Wilson, P. G. Rutaceae. In *Flowering Plants. Eudicots*; Springer: Berlin, Heidelberg, 2010; pp 276–356.
2. Groppo, M.; Pirani, J. R.; Salatino, M. L.; Blanco, S. R.; Kallunki, J. A. Phylogeny of Rutaceae based on two noncoding regions from cpDNA. *Am. J. Bot.* **2008,** *95* (8), 985–1005.
3. Gray, A. I.; Waterman, P. G. Coumarins in the Rutaceae. *Phytochemistry* **1978,** *17* (5), 845–864.
4. Appelhans, M. S.; Reichelt, N.; Groppo, M.; Paetzold, C.; Wen, J. Phylogeny and Biogeography of the Pantropical Genus Zanthoxylum and Its Closest Relatives in the Proto-Rutaceae Group (Rutaceae). *Mol. Phylogenet. Evol.* **2018,** *126*, 31–44.
5. Appelhans, M. S.; Wen, J. Phylogenetic Placement of Ivodea and Biogeographic Affinities of Malagasy Rutaceae. *Plant Syst. Evol.* **2020,** *306* (1), 1–14.
6. Ombito, J. O.; Chi, G. F.; Wansi, J. D. Ethnomedicinal Uses, Phytochemistry, and Pharmacology of the Genus Vepris (Rutaceae): A Review. *J. Ethnopharmacol.* **2020,** 113622.
7. Liaqat, I.; Riaz, N.; Saleem, Q. U. A.; Tahir, H. M.; Arshad, M.; Arshad, N. Toxicological Evaluation of Essential Oils from Some Plants of Rutaceae Family. *Evid. Based Complement. Alternat. Med.* **2018,** *2018*.
8. Kaigongi, M. M.; Lukhoba, C. W.; Yaouba, S.; Makunga, N. P.; Githiomi, J.; Yenesew, A. In Vitro Antimicrobial and Antiproliferative Activities of the Root Bark Extract and Isolated Chemical Constituents of Zanthoxylum Paracanthumkokwaro (Rutaceae). *Plants* **2020,** *9* (7), 920.
9. Adeola, H. A.; Bano, A.; Vats, R.; Vashishtha, A.; Verma, D.; Kaushik, D.; ... Bhardwaj, R. Bioactive Compounds and Their Libraries: An Insight into Prospective Phytotherapeutics Approach for Oral Mucocutaneous Cancers. *Biomed. Pharm.* **2021,** *141*, 111809.
10. Paetzold, C.; Wood, K. R.; Eaton, D. A.; Wagner, W. L.; Appelhans, M. S. Phylogeny of Hawaiian Melicope (Rutaceae): RAD-seq Resolves Species Relationships and Reveals Ancient Introgression. *Front. Plant Sci.* **2019,** *10*, 1074.
11. Sun, K.; Liu, Q. Y.; Wang, A.; Gao, Y. W.; Zhao, L. C.; Guan, W. B. Comparative Analysis and Phylogenetic Implications of Plastomes of Five Genera in Subfamily Amyridoideae (Rutaceae). *Forests* **2021,** *12* (3), 277.
12. Paetzold, C.; Wood, K. R.; Eaton, D. A.; Wagner, W. L.; Appelhans, M. S. Phylogeny of Hawaiian Melicope (Rutaceae): RAD-seq Resolves Species Relationships and Reveals Ancient Introgression. *Front Plant Sci.* **2019,** *10*, 1074.
13. Tian-chi, G.; Yin-min, Y. A New Species of Citrus (Rutaceae) from China. *J. Syst. Evol.* **1997,** *35* (4), 353–355.
14. Nagano, Y.; Mimura, T.; Kotoda, N.; Matsumoto, R.; Nagano, A. J.; Honjo, M. N.; ... Yamamoto, M. Phylogenetic Relationships of Aurantioideae (Rutaceae) Based on RAD-Seq. *Tree Genet. Genomes* **2018,** *14* (1), 6.
15. Zhang, M.; Wang, J.; Zhu, L.; Li, T.; Jiang, W.; Zhou, J.; ... Wu, C. Zanthoxylum Bungeanum Maxim. (Rutaceae): A Systematic Review of Its Traditional Uses,

Botany, Phytochemistry, Pharmacology, Pharmacokinetics, and Toxicology. *Int. J. Mol. Sci.* **2017,** *18* (10), 2172.
16. Saraiva Filho, D. E.; Sousa, J. B. D.; Santos, H. S. D.; Fontenelle, R. O. D. S. Chemical Compounds Isolated from Extracts and Essential Oils of the Genus Zanthoxylum Linnaeus (Rutaceae) and Its Antimicrobial Potential. *Hoehnea* **2020,** *47*.
17. Zhang, W.; Tan, S.; Xi, W.; Yang, J.; Liao, Q.; Lan, J.; ... Tang, J. Comparison of Volatile Components in Fresh and Dried Zanthoxylum Bungeanum Maxim. *Food Sci. Biotechnol.* **2019,** *28* (4), 1083–1092.
18. Sun, L.; Yu, D.; Wu, Z.; Wang, C.; Yu, L.; Wei, A.; Wang, D. Comparative Transcriptome Analysis and Expression of Genes Reveal the Biosynthesis and Accumulation Patterns of Key Flavonoids in Different Varieties of Zanthoxylum Bungeanum Leaves. *J. Agri. Food. Chem.* **2019,** *67* (48), 13258–13268.
19. Chen, X.; Wei, Z.; Zhu, L.; Yuan, X.; Wei, D.; Peng, W.; Wu, C. Efficient Approach for the Extraction and Identification of Red Pigment from Zanthoxylum Bungeanum Maxim and Its Antioxidant Activity. *Molecules* **2018,** *23* (5), 1109.
20. Yang, Q.; Mei, X.; Wang, Z.; Chen, X.; Zhang, R.; Chen, Q.; Kan, J. Comprehensive Identification of Non-volatile Bitter-tasting Compounds in Zanthoxylum Bungeanum Maxim. by Untargeted Metabolomics Combined with Sensory-guided Fractionation Technique. *Food Chem.* **2021,** *347*, 129085.
21. Chen, X.; Wang, W.; Wang, C.; Liu, Z.; Sun, Q.; Wang, D. Quality Evaluation and Chemometric Discrimination of Zanthoxylum Bungeanum Maxim Leaves Based on Flavonoids Profiles, Bioactivity and HPLC-fingerprint in a Common Garden Experiment. *Ind. Crops Prod.* **2019,** *134*, 225–233.
22. Fan, L.; Huang, Y.; Zhao, R.; Fan, W.; Zhang, M.; Zhang, H.; ... Wu, C. Geographical-origin Discrimination and Volatile Oil Quantitative Analysis of Zanthoxylum Bungeanum Maxim. with a Portable Near-infrared Spectrometer. *Analyt. Methods* **2019,** *11* (41), 5301–5310.
23. Shi, J.; Fei, X.; Hu, Y.; Liu, Y.; Wei, A. Identification of Key Genes in the Synthesis Pathway of Volatile Terpenoids in Fruit of Zanthoxylum Bungeanum Maxim. *Forests* **2019,** *10* (4), 328.
24. Fu, L.; Xie, H.; Shi, S. Multielement Analysis of Zanthoxylum Bungeanum Maxim. Essential Oil Using ICP-MS/MS. *Analyt. Bioanalyt. Chem.* **2018,** *410* (16), 3769–3778.
25. Zhao, M.; Tang, X.; Gong, D.; Xia, P.; Wang, F.; Xu, S. Bungeanum Improves Cognitive Dysfunction and Neurological Deficits in D-galactose-induced Aging Mice via Activating PI3K/Akt/Nrf2 Signaling Pathway. *Front Pharmacol.* **2020,** *11*, 71.
26. Guo, T.; Liu, Y. L.; Song, T. T.; Xin, H. L.; Chang, J.; Qin, L. P. Water-Soluble Constituents of Zanthoxylum Bungeanum. *Chem. Nat. Compd.* **2020,** *56* (1), 145–146.
27. Yin, X.; Xu, X.; Zhang, Q.; Xu, J. Rapid Determination of the Geographical Origin of Chinese Red Peppers (Zanthoxylum Bungeanum Maxim.) Based on Sensory Characteristics and Chemometric Techniques. *Molecules* **2018,** *23* (5), 1001.
28. Tao, X.; Peng, W.; Xie, D.; Zhao, C.; Wu, C. Quality Evaluation of Hanyuan Zanthoxylum Bungeanum Maxim. Using Computer Vision System Combined with Artificial Neural Network: A Novel Method. *Int. J. Food Prop.* **2017,** *20* (12), 3056–3063.

29. Liu, M.; Liu, J.; Atzberger, C.; Jiang, Y.; Ma, M.; Wang, X. Zanthoxylum bungeanum Maxim Mapping with Multi-temporal Sentinel-2 Images: The Importance of Different Features and Consistency of Results. *ISPRS J. Photogramm. Remote Sens.* **2021,** *174*, 68–86.
30. Pang, W.; Liu, S.; He, F.; Li, X.; Saira, B.; Zheng, T.; ... Pei, X. F. Anticancer Activities of Zanthoxylum bungeanum Seed Oil on Malignant Melanoma. *J. Ethnopharmacol.* **2019,** *229*, 180–189.
31. Sun, X.; Zhang, D.; Zhao, L.; Shi, B.; Xiao, J.; Liu, X.; ... Zou, X. Antagonistic Interaction of Phenols and Alkaloids in Sichuan Pepper (Zanthoxylum bungeanum) Pericarp. *Ind. Crops Prod.* **2020,** *152*, 112551.
32. Zhao, M.; Dai, Y.; Li, P.; Wang, J.; Ma, T.; Xu, S. Inhibition of NLRP3 Inflammasome Activation and Pyroptosis with the Ethyl Acetate Fraction of Bungeanum Ameliorated Cognitive Dysfunction in Aged Mice. *Food Funct.* **2021.**
33. Kasimu, N.; Yu, D.; Sun, L.; Zheng, R.; Wang, D. Development of 1H-NMR Methods for Quantitative Determination of Alkylamides in Zanthoxylum Bungeanum and Quality Evaluation based on its Fingerprint. *J. Food Sci.* **2021,** *86* (9), 3951–3963.
34. Feng, S.; Liu, Z.; Cheng, J.; Li, Z.; Tian, L.; Liu, M.; ... Wei, A. Zanthoxylum-specific Whole Genome Duplication and Recent Activity of Transposable Elements in the Highly Repetitive Paleotetraploid Z. Bungeanum Genome. *Horticulture Res.* **2021,** *8* (1), 1–15.
35. Ombito, J. O.; Chi, G. F.; Wansi, J. D. Ethnomedicinal Uses, Phytochemistry, and Pharmacology of the Genus Vepris (Rutaceae): A Review. *J. Ethnopharmacol* **2020,** 113622.
36. Morton, C. M. Phylogenetic Relationships of Vepris (Rutaceae) Inferred from Chloroplast, Nuclear, and Morphological Data. *PLoS One* **2017,** *12* (3), e0172708.
37. Cheek, M.; Gosline, G.; Onana, J. M. Veprisbali (Rutaceae), a New Critically Endangered (Possibly extinct) Cloud Forest Tree Species from Bali Ngemba, Cameroon. *Willdenowia* **2018,** *48* (2), 285–292.
38. Lachenaud, O.; Onana, J. M. The West and Central African Species of Vepris Comm. ex A. Juss. (Rutaceae) with Simple or Unifoliolate Leaves, Including Two New Combinations. *Adansonia* **2021,** *43* (10), 107–116.
39. Onana, J. M.; Cheek, M.; Chevillotte, H. Additions au Genre Vepris Comm. ex A. Juss. (Rutaceae-Toddalieae) au Cameroun. *Adansonia* **2019,** *41* (1), 41–52.
40. Cheek, M.; Onana, J. M.; Yasuda, S.; Lawrence, P.; Ameka, G.; Buinovskaja, G. Addressing the Veprisverdoorniana Complex (Rutaceae) in West Africa, with Two New Species. *Kew. Bull.* **2019,** *74* (4), 1–16.
41. Langat, M.; Kami, T.; Cheek, M. Chemistry, Taxonomy and Ecology of the Potentially Chimpanzee-dispersed Vepristeva sp. nov. (Rutaceae) of Coastal Thicket in the Congo Republic. *bioRxiv* **2021.**
42. Kouam, A. D. K.; Bissoue, A. N.; Tcho, A. T.; Happi, E. N.; Waffo, A. F. K.; Sewald, N.; Wansi, J. D. Antimicrobial Furoquinoline Alkaloids from Veprislecomteana (Pierre) Cheek T. Heller (Rutaceae). *Molecules* **2018,** *23* (1), 13.
43. Noulala, C. G. T.; Fotso, G. W.; Rennert, R.; Lenta, B. N.; Sewald, N.; Arnold, N.; ... Ngadjui, B. T. Mesomeric Form of Quaternary Indoloquinazoline Alkaloid and Other Constituents from the Cameroonian Rutaceae Araliopsissoyauxii Engl. *Biochem Syst Ecol.* **2020,** *91*, 104050.

44. Kouam, A. D. K.; Kenmogne, S. B.; Lobe, J. S.; Happi, E. N.; Stammler, H. G.; Waffo, A. F. K.; ... Wansi, J. D. A Rotamerictryptamide Alkaloid from the Roots of Veprislecomteana (Pierre) Cheek & T. Heller (Rutaceae). *Fitoterapia* **2019,** *135*, 9–14.
45. Fajinmi, O. O.; Kulkarni, M. G.; Benická, S.; Zeljković, S. Ć.; Doležal, K.; Tarkowski, P.; ... Van Staden, J. Antifungal activity of the volatiles of Agathosmabetulina and Coleonema Album Commercial Essential Oil and Their Effect on the Morphology of Fungal Strains Trichophyton Rubrum and T. Mentagrophytes. *S. Afr. J. Bot.* **2019,** *122*, 492–497.
46. Witbooi, H.; Okem, A.; Makunga, N. P.; Kambizi, L. Micropropagation and Secondary Metabolites in Agathosmabetulina (Berg.). *S. Afr. J. Bot.* **2017,** *111*, 283–290.
47. Fajinmi, O. O.; Olarewaju, O. O.; Van Staden, J. Traditional Use of Medicinal and Aromatic Plants in Africa. In *Medicinal and Aromatic Plants of the World-Africa Volume 3*; Springer, Dordrecht, 2017; pp 61–76.
48. AL-Zuaidy, M. H.; Ismail, A.; Mohamed, S.; Razis, A. F. A.; Mumtaz, M. W.; Hamid, A. A. Antioxidant Effect, Glucose Uptake Activity in Cell Lines and Cytotoxic Potential of Melicopelunu-ankenda Leaf Extract. *J. Herb. Med.* **2018,** *14*, 55–60.
49. Van Vuuren, S.; Ramburrun, S.; Kamatou, G.; Viljoen, A. Indigenous South African Essential Oils as Potential Antimicrobials to Treat Foot Odour (Bromodosis). *S. Afr. J. Bot.* **2019,** *126*, 354–361.
50. Campos Portela, R.; de Sousa Fiuza, T.; Realino de Paula, J.; Luiz Borges, L. Phytochemistry, Pharmacology and Botanical Aspects of Hortia Species in the Search for Molecules with Bioactive Potential-A Review. *Pharmacognosy Rev.* **2020,** *14* (28).
51. Ndlovu, B.; Africa, C.; Klaasen, J.; Rahiman, F. The Use of South African Medicinal Plants in the Treatment of Fungal Skin Infections: A Systematic Review. *Australas Med. J. (Online)* **2019,** *12* (8), 246–254.
52. Abe, B. T.; Jordaan, J. Separability Method for Homogeneous Leaves Using Spectroscopic Imagery and Machine Learning Algorithms. In *International Symposium on Neural Networks*; Springer, Cham, 2019, July; pp 282–291.
53. Abe, B.; Jordaan, J. In *Classification of Agathosma Plant Species from Overlapping Leaves Imagery*, Iaeng Transactions on Engineering Sciences: Special Issue for the International Association of Engineers Conferences 2016; 2018; pp 489–499.
54. Roberts, R.; Lin, H.; Pietersen, G. Genetic Diversity of 'Candidatus Liberibacter Africanus' in South Africa Based on Microsatellite Markers. *Eur. J. Plant Pathol.* **2021,** *159* (2), 259–268.
55. Stohs, S. J. Safety, efficacy, and Mechanistic Studies Regarding Citrus Aurantium (Bitter Orange) Extract and p-synephrine. *Phytother. Res.* **2017,** *31* (10), 1463–1474.
56. Shara, M.; Stohs, S. J.; Smadi, M. M. Safety Evaluation of p-synephrine following 15 days of Oral Administration to Healthy Subjects: A Clinical Study. *Phytother. Res.* **2018,** *32* (1), 125–131.
57. Stohs, S. J.; Shara, M.; Ray, S. D. p-Synephrine, Ephedrine, p-octopamine and m-synephrine: Comparative Mechanistic, Physiological and Pharmacological Properties. *Phytother. Res.* **2020,** *34* (8), 1838–1846.
58. Stohs, S. J.; Ray, S. D. Review of Published Bitter Orange Extract and p-synephrine Adverse Event Clinical Study Case Reports. *J. Dietary Suppl.* **2020,** *17* (3), 355–363.

59. Ruiz-Moreno, C.; Del Coso, J.; Giráldez-Costas, V.; González-García, J.; Gutiérrez-Hellín, J. Effects of p-Synephrine During Exercise: A Brief Narrative Review. *Nutrients* **2021,** *13* (1), 233.
60. Takagi, M.; Kimura, K.; Nakashima, K. I.; Hirai, T.; Inoue, M. Induction of Beige Adipocytes by Naturally Occurring β3-adrenoceptor Agonist p-synephrine. *Eur. J. Pharmacol.* **2018,** *836*, 67–74.
61. Suntar, I.; Khan, H.; Patel, S.; Celano, R.; Rastrelli, L. An Overview on Citrus Aurantium L.: Its Functions as Food Ingredient and Therapeutic Agent. *Oxidative Med. Cell Longevity* **2018,** *2018*.
62. Stohs, S. J.; Ratamess, N. A. Effects of p-synephrine in Combination with Caffeine: A Review. *Nutrit. Dietary Suppl.* **2017,** *9*, 87–96.
63. Koh, A. H. W.; Chess-Williams, R.; Lohning, A. E. Racemic Synephrine found in Citrus Aurantium-listing Pre-workout Supplements Suggests a Non-plant-based Origin. *Drug Test Anal.* **2021.**
64. da Silva-Pereira, J. F.; de Oliveira Valoto, A. L.; Bracht, L.; de Almeida Gonçalves, G.; Peralta, R. M.; Bracht, A. The Action of p-synephrine on Lipid Metabolism in the Perfused Rat Liver. *J. Biosci. Med.* **2017,** *5* (5), 8–21.
65. Odeyemi, O. O.; Yakubu, M. T.; Masika, P. J.; Afolayan, A. J. Toxicological Evaluation of the Essential Oil from Mentha Longifolia L. Subsp. Capensis Leaves in Rats. *J. Med. Food* **2009,** *12* (3), 669–674.
66. Portela García, G. Control of the Quality of the Raw Material Obtained from the Bark of Citrus Sinensis L, in the Local Production Centers of the Province of Villa Clara. Doctoral Dissertation, Universidad Central Marta Abreu de Las Villas. Faculty of Chemistry and Pharmacy. Department Degree in Pharmaceutical Sciences, 2019.
67. Vinturelle, R.; Mattos, C.; Meloni, J.; Nogueira, J.; Nunes, M. J.; Vaz, I. S.; ... Chagas, E. F. D. In Vitro Evaluation of Essential Oils Derived from Piper nigrum (Piperaceae) and Citrus limonum (Rutaceae) against the Tick Rhipicephalus (Boophilus) Microplus (Acari: Ixodidae). *Biochem. Res. Int.* **2017,** *2017*.
68. Tan, L. P.; Hamdan, R. H.; Hassan, B. N. H.; Reduan, M. F. H.; Okene, I. A. A.; Loong, S. K.; ... Lee, S. H. Rhipicephalus Tick: A Contextual Review for Southeast Asia. *Pathogens* **2021,** *10* (7), 821.
69. Suwannayod, S.; Sukontason, K. L.; Pitasawat, B.; Junkum, A.; Limsopatham, K.; Jones, M. K.; ... Sukontason, K. Synergistic Toxicity of Plant Essential Oils Combined with Pyrethroid Insecticides Against Blow Flies and the House Fly. *Insects* **2019,** *10* (6), 178.
70. Pandey, L.; Upadhyay, R. K. Anti-Termite Efficacy of Various Plant Essential Oils with Special Reference to Family Rutaceae, 2021.
71. Guettal, S.; Tine, S.; Tine-Djebbar, F.; Soltani, N. Evaluation of Citrus limonum (Sapindales: Rutaceae) L. Essential Oil as Protectant Against the Granary Weevil, Sitophilus granarius (L.)(Coleoptera: Curculionidae). *Allelopathy J.* **2020,** *51* (1), 79–92.
72. Guo, S. S.; Wang, Y.; Chen, Z. Y.; Zhang, Z.; Cao, J. Q.; Pang, X.; ... Du, S. S. Essential Oils from Clausena species in China: Santalene Sesquiterpenes Resource and Toxicity Against Liposcelis Bostrychophila. *J. Chem.* **2018,** *2018*.
73. De Souza, M. A.; Da Silva, L.; Macêdo, M. J. F.; Lacerda-Neto, L. J.; dos Santos, M. A. C.; Coutinho, H. D. M.; Cunha, F. A. B. Adulticide and Repellent Activity of

Essential Oils Against Aedes Aegypti (Diptera: Culicidae)–A Review. *S. Afr. J. Bot.* **2019,** *124,* 160–165.
74. Elshafie, H. S.; Camele, I. An Overview of the Biological Effects of Some Mediterranean Essential Oils on Human Health. *BioMed. Res. Int.* **2017,** *2017.*
75. Elshafie, H. S.; Mancini, E.; Sakr, S.; De Martino, L.; Mattia, C. A.; De Feo, V.; Camele, I. Antifungal Activity of Some Constituents of Origanum Vulgare L. Essential Oil Against Postharvest Disease of Peach Fruit. *J. Med. Food.* **2015,** *18* (8), 929–934.
76. Bhattacharya, T.; Dutta, S.; Akter, R.; Rahman, M.; Karthika, C.; Nagaswarupa, H. P.; ... Bungau, S. Role of Phytonutrients in Nutrigenetics and Nutrigenomic Perspective in Curing Breast Cancer. *Biomolecules* **2021,** *11* (8), 1176.
77. Walia, V.; Kaushik, D.; Mittal, V.; Kumar, K.; Verma, R.; Parashar, J.; ... Ashraf, G. M. Delineation of Neuroprotective Effects and Possible Benefits of AntioxidantsTherapy for the Treatment of Alzheimer's Diseases by Targeting Mitochondrial-Derived Reactive Oxygen Species: Bench to Bedside. *Mol. Neurobiol.* **2021,** 1–24.
78. Chopra, H.; Dey, P. S.; Das, D.; Bhattacharya, T.; Shah, M.; Mubin, S.; ... Alamri, B. M. Curcumin Nanoparticles as Promising Therapeutic Agents for Drug Targets. *Molecules* **2021,** *26* (16), 4998.
79. Karthika, C.; Hari, B.; Rahman, M. H.; Akter, R.; Najda, A.; Albadrani, G. M.; ... Abdel-Daim, M. M. Multiple Strategies with the Synergistic Approach for Addressing Colorectal Cancer. *Biomed. Pharmacother.* **2021,** *140,* 111704.
80. Karthika, C.; Appu, A. P.; Akter, R.; Rahman, M.; Tagde, P.; Ashraf, G. M.; ... Bungau, S. Potential Innovation Against Alzheimer's Disorder: a Tricomponent Combination of Natural Antioxidants (Vitamin E, Quercetin, and Basil Oil) and the Development of Its Intranasal Delivery. *Environ. Sci. Pollut. Res.* **2022,** 1–16.
81. Arya, A.; Chahal, R.; Nanda, A.; Kaushik, D.; Bin-Jumah, M.; Rahman, H.; ... Mittal, V. Statistically Designed Extraction of Herbs Using Ultrasound Waves: A Review. *Curr. Pharm. Des.* **2021.**
82. Akter, R.; Rahman, M.; Bhattacharya, T.; Kaushik, D.; Mittal, V.; Parashar, J.; ... Tagde, P. Novel Coronavirus Pathogen in Humans and Animals: An Overview on Its Social Impact, Economic Impact, and Potential Treatments. *Environ. Sci. Pollut. Res.* **2021,** 1–19.
83. Rahman, M.; Bajgai, J.; Fadriquela, A.; Sharma, S.; Trinh, T. T.; Akter, R.; ... Lee, K. J. Therapeutic Potential of Natural Products in Treating Neurodegenerative Disorders and Their Future Prospects and Challenges. *Molecules* **2021,** *26* (17), 5327.
84. Akter, R.; Rahman, M.; Kaushik, D.; Mittal, V.; Uivarosan, D.; Nechifor, A. C.; ... Bungau, S. Chemo-Preventive Action of Resveratrol: Suppression of p53—A Molecular Targeting Approach. *Molecules* **2021,** *26* (17), 5325.
85. Bhattacharya, T.; Dey, P. S.; Akter, R.; Kabir, M. T.; Rahman, M. H.; Rauf, A. Effect of Natural Leaf Extracts as Phytomedicine in Curing Geriatrics. *Experiment Gerontol.* **2021,** *150,* 111352.

CHAPTER 11

Antioxidant Properties of the Family Rutaceae

ANEES AHMED KHALIL[1], AMMAR AHMAD KHAN[2], MUHAMMAD ARSLAN KHAN[3], and SAIMA NAZ[4]

[1]University Institute of Diet and Nutritional Sciences, Faculty of Allied Health Sciences, The University of Lahore, Lahore, Pakistan

[2]University Institute of Food Science and Technology, Faculty of Allied Health Sciences, The University of Lahore, Lahore, Pakistan

[3]Department of Pharmacy, The University of Lahore, Lahore, Pakistan

[4]Department of Biotechnology, Bacha Khan University Charsadda, Khyber Pakhtunkhwa (KP), Pakistan

ABSTRACT

Plant-based natural phytochemicals with significant antioxidant properties have attained the focus of researchers owing to their roles in disease prevention and treatment. Nowadays, consumers prefer natural antioxidants as compared to those of synthetic origin owing to emotional reasons and personal preferences. The family Rutaceae, comprising the various renowned genus, has evolved as a significant plant family extensively studied in recent times. Species present in the family Rutaceae have been a source of traditional medicine in Ayurvedic and Unani medication systems. Currently, extracts and isolates from different parts (flowers, fruits, stem barks, roots, and leaves) of members of Rutaceae are being evaluated scientifically for

Phytochemical and Pharmacological Investigation of the Family Rutaceae.
Abdur Rauf, Hafiz Ansar Rasul Suleria, & Syed Muhammad Mukarram Shah (Eds.)

their biological activities. Phytochemical screening has shown that these parts are rich sources of various secondary metabolites with significant antioxidant properties. Numerous in vitro, in vivo, and clinical trials have validated that Rutaceae plant extracts from different parts possess various health benefits owing to the presence of antioxidants. This book chapter highlights the potent antioxidant properties of a few renowned members (*Aegle*, *Atalantia*, *Citrus*, and *Clausena*) of the family Rutaceae.

11.1 INTRODUCTION

In recent times, scientists are focusing more on exploring plant-based natural antioxidants owing to their associated health-promoting benefits. Medicinal plants are known to have various antioxidant constituents that help in scavenging free radicals and reactive oxygen species.[24] Reactive oxygen species including different free radicals like hydroxyl radicals ($OH^{\bullet}$), superoxide anion radicals ($O_2^{\bullet}(-)$), and nonradical species (singlet oxygen and H_2O_2) are numerous forms of activated oxygen. All of these species are considered as aggravating elements in the aging process and cellular injuries.[13] Sugars, DNA molecules, lipids, RNA molecules, and proteins are known as the main targets of RSS (reactive sulfur species), RNS (reactive nitrogen species), and ROS.[7,24]

Generation of free radicals is a continuous process in vivo, but when produced in excess, they cause oxidative stress in pathological conditions.[23] These free radicals are highly unstable and have the capacity to damage various organic molecules like carbohydrates, proteins, and DNA. Several ailments like neurological disorders, aging, inflammation, cancer, cardiovascular diseases, and autoimmune disorders are linked to oxidative stress.[25,49] Medicinal plants have the ability to produce ample amounts of secondary metabolites having diverse structural properties.[5] Phytochemical screening of secondary metabolites from medicinal plants to assess their pharmacological properties has been reported to be an immense source of countless beneficial potentials demonstrating molecular-based diversity bioengineered by nature. Consumption of plant-based natural antioxidants has been reported to be helpful in protecting macromolecules from oxidative damages.[36] Antioxidants, both natural and synthetic, are referred to compounds having the potential to scavenge free radical, inhibit lipid peroxidation, and chelate ROS.[21] Natural antioxidants are constituents having the capability to either inhibit or delay lipid oxidation and/or other molecules through the inhibition of propagating and initiating oxidizing chain reactions.[50] Plant extracts are a cache of antioxidants having the ability to avert oxidation and its related damages.

The Rutaceae family includes nearly 1600 species of shrubs, trees, and flowering plants thoroughly distributed across diverse regions of the world, specifically tropical and temperate regions.[30] The main and most renowned genera of the family Rutaceae are *Aegle*, *Atalantia*, *Citrus*, and *Clausena*.[40] Mainly, species of the family Rutaceae comprise aromatic plants with volatile aromatic compounds in different parts of plants, such as fruits, seeds, and leaves.[2] Since ancient times, various species of the family Rutaceae are being used for manufacturing perfumes, functional foods, and folklore medicines. Additionally, it is evident from the literature that the family Rutaceae is a rich source of plant secondary metabolites with antioxidant properties that help in reducing the disease burden. Phytochemical studies have shown presence of tannins, coumarins, essential oils, alkaloids, flavonoids, and other phenolics possessing antioxidant properties.[22] The naturally present antioxidants in the family Rutaceae possess diverse bioactivities, including antiprotozoal, antidiarrheal, antimicrobial, antioxidant, and anticancer properties.[44] Keeping in view the importance of antioxidants and their presence in the family Rutaceae, this chapter focuses on the antioxidant properties associated with some renowned genera (*Aegle*, *Atalantia*, *Citrus*, and *Clausena*) of the family Rutaceae.

11.2 ANTIOXIDANT PROPERTIES OF GENUS AEGLE

As already discussed in Chapter 1, genus *Aegle* comprises three accepted species, out of which most work has been conducted on the species *Aegle marmelos* (L.) Corrêa. *Aegle marmelos* possesses various metabolites that have shown therapeutic effects as reported by several scientists in different studies. Evidently, different parts of *Aegle marmelos* contain a diverse class of constituents, like skimmianine, aegelenine, marmelosin, aeglin, imperatorin, and marmesin, that have been successfully isolated from plant matrices and studied extensively for their bioactivities. In ethnomedicine, *A. marmelos* has been explored for its bioactivities including antioxidant, anticancer, anti-inflammatory, and antidiabetic properties.[37] The following section highlights the antioxidant properties of genus *Aegle*.

In Ayurvedic medicine system, Bael (*A. marmelos*) has played an important role in promoting consumer health owing to their therapeutic and pharmacological properties. In recent times, Bael has attained attention of scientists owing to its antioxidant properties.[9,28] Further, various reviews have been published signifying the health-promoting benefits of Bael in modern therapeutics.[51] Nevertheless, clinical trials are suggested to validate the in vitro bioactivities associated with the plant Bael. Accordingly, a study conducted by Rautela et al.[34] revealed the antioxidative properties of *A. marmelos* leaves in cows having endometritis. The study comprises two groups, a control group and a treated group, each consisting of six cows. In the treatment group, each cow was administered with *A. marmelos* leaf powder (200 g). Administration of *A. marmelos* leaf powder resulted in a reduction of malondialdehyde content from 1.29 (0 day) to 0.36 (10 days) nmol/mL. However, the concentration of antioxidant enzyme superoxide dismutase and serum ascorbic acid content increased due to the administration of *A. marmelos* leaf powder. Further, the serum total antioxidant capacity also increased in *A. marmelos* leaf-treated group with time from day 0 (0.65 mmol/L) to day 10 (0.86 mmol/L). As compared to control group, the content of glutathione reduced significantly in cows supplemented with *A. marmelos* leaf powder.[34] Table 11.1 shows the antioxidant properties of *Aegle marmelos* (L.) as reported by different scientists.

A. marmelos, also known as "stone apple," has traditionally been used for its nutritional and medicinal properties, but still is considered to be an underutilized plant owing to limited knowledge regarding its nutritional and phytochemical composition. This plant is native to subcontinent and is used extensively in summer season. Bael/stone apple is considered to be a rich source of dietary fiber, protein, and minerals. Different parts of this plant have been used for treating different diseases such as inflammation, cancer, diabetes, aging, oxidation, and gastrointestinal problems.[42] Pectin extracted from different parts of Bael, such as seed, pulp, and shell, has been investigated for its antioxidative properties. Pectin extracted from pulp showed maximum (IC_{50}: 10.60 μg/mL) antioxidant activity as compared to pectin extracted from seed and shell (i.e., IC_{50}: 33.47 and IC_{50}: 53.08 μg/mL, respectively). Likewise, pectin extracted from the pulp of Bael possessed highest total phenolics (43.50 mgGAE/g) as compared to pectin extracted from seed (33.5 mgGAE/g) and shell (23.50 mgGAE/g) of bael. The outcomes revealed that different examined parts of bael fruit showed good-quality pectin, and therefore could be utilized as a nutraceutical compound.[45]

A. marmelos, abundantly found in sub-Himalayan regions, is a rich source of antioxidants and phenolics. A study investigated the phytochemical screening of bael dried leaf extract and reported that it is a source of tannins, alkaloids, saponins, flavonoids, steroids, and phenols. The antioxidant activity (AA) of bael leaf extracts examined using DPPH assay revealed water extract having the highest inhibition, that is, 79%.[48] Ethanolic bael leaf extract demonstrated total phenolic contents as 218.33 ± 43.81 mgGAE/g of dry extract, whereas the reported antioxidant capacity was 33.62 ± 3.07 mgTEAC/g of dry extract. Bael leaf extract at a minimum concentration also revealed a good antioxidant potential owing to their ability to reduce the content of malondialdehyde and reactive oxygen species in HepG2 cells. Antioxidant potential associated with bael leaf can be beneficial in developing new therapeutics. Nevertheless, further investigations are required to properly understand the mechanism underlying its health-promoting benefits.[18]

Inflammation and oxidative stress are considered crucial in implicating various lifestyle-related syndromes like cancer, ulcer, cardiovascular problems, and diabetes.[32] Accordingly, a study isolated and characterized furanocoumarin from bael fruits, which was reported to have an effective modulating effect on inflammation and oxidative stress. Extract (ethyl acetate) of bael fruit was found to be predominant in marmelosin. Marmelosin isolated from bael fruit possessed potential antioxidative and antiproliferative characteristics with a reported IC_{50} of 15.40 ± 0.32 μM and 6.24 ± 0.16 μM, respectively. Marmelosin showed potent antioxidative, anti-inflammatory, and anticancer characteristics through TNF-α-mediated Akt signaling pathway.[32] Likewise, *A. marmelos* leaf extract has been found to increase the content of glutathione peroxidase, superoxide dismutase, and catalase in liver and blood of rats that were administered with the extract.[38] Various parts (stem bark, roots, and leaves) of *A. marmelos* have been examined to assess their comparative antioxidant properties. For this purpose, a study was conducted revealing total phenolic and total flavonoid contents varying from 1.72 to 9.83 mg/g and 1.08 to 8.24 mg/g, respectively. The free radical scavenging property of the experimented extracts calculated using the DPPH assay was noticed to be the maximum in methanolic leaf extract, that is, IC_{50}: 2.096 μg/mL. The antioxidant activity of bael leaf extract was observed to be 10 times higher as compared to BHT (butylated hydroxyl toluene). This prominent antioxidant property of the bael leaf was due to the presence of high amounts of phenolic compounds in its methanolic extracts.[40]

Human beings, for their easing effect against gastrointestinal problems, have traditionally utilized bael fruit pulp. Evidently, this property is linked with antioxidant property of phenols present in its fruit pulp. In a study, bael fruit pulp extract was subjected to standard procedures for screening its antioxidant properties and identifying phytochemicals. Results of this study revealed that the aqueous extract of bael fruit possesses an array of lignin, steroids, flavonoids, saponins, terpenoids, and tannins, while alcoholic extract had alkaloids. Further, in vitro examination of antioxidant properties of bael fruit extract showed that both aqueous (IC_{50}: 37.11–158.99 μg/mL) and alcoholic (35.02–283.06 μg/mL) extracts possessed strong antioxidant power.[33] Likewise, Reddy and Urooj[35] compared the antioxidant properties of bael leaves extracted using water, ethanol, and methanol as extracting solvents. They determined the antioxidant power using DPPH and FRAP assays. They were of the view that water extract of bael leaves possessed maximum (93%) radical scavenging activity as compared to ethanolic (87%) and methanolic (81%) extracts. Conclusively, they reported that bael leaves have the potential to be used as an alternative/natural source of antioxidants.[35]

TABLE 11.1 Antioxidant Properties of *Aegle Marmelos* (L.).

Parameters	Plant parts	Results	References
Serum total antioxidant capacity	Leaf powder	0.86 mmol/L	34
DPPH: Antioxidant activity	Pulp pectin	IC_{50}: 10.60 μg/mL	45
	Seed pectin	IC_{50}: 33.47 μg/mL	
	Shell pectin	IC_{50}: 53.08 μg/mL	
Total phenolics	Pulp pectin	43.50 mgGAE/g	
	Seed pectin	33.5 mgGAE/g	
	Shell pectin	23.50 mgGAE/g	
DPPH: Antioxidant activity	Leaf water extract	Percent inhibition: 79%	48
Total phenolics contents	Ethanolic bael leaf extract	218.33 mgGAE/g	18
ABTS radical cation-scavenging activity	Ethanolic bael leaf extract	33.62 mgTEAC/g	
DPPH: Antioxidant activity	Ethyl acetate bael fruit extract	IC_{50}: 15.40 μM	32
DPPH: Antioxidant activity	Methanolic bael leaf extract	IC_{50}: 2.096 μg/mL	40

TABLE 11.1 *(Continued)*

Parameters	Plant parts	Results	References
DPPH: Antioxidant activity	Aqueous bael fruit extract	IC_{50}: 37.11–158.99 µg/mL	33
	Alcoholic bael fruit extract	IC_{50}: 35.02–283.06 µg/mL	
DPPH: Antioxidant activity	Bael leaves aqueous extract	93%	35
	Bael leaves ethanolic extract	87%	
	Bael leaves methanolic extract	81%	

11.3 ANTIOXIDANT PROPERTIES OF GENUS ATALANTIA

Various species of the genus *Atalantia* have been extensively used traditionally and in Ayurvedic medication systems. Among these, the most consumed species in India are *A. racemose*, *A. monophylla*, and *A. wightii.* The fruits of these species have been widely consumed by tribal people. Numerous studies have been conducted to assess the antioxidant properties of *A. wightii*, *A. monophylla*, and *A. racemose*. Purposely, DPPH assay was performed to assess the antioxidative properties of these three species of the genus *Atalantia.* Results of the study revealed IC_{50} values of 348.75, 375.64, and 336.84 µg mL^{-1} for *A. racemose*, *A. wightii*, and *A. monophylla*, respectively as compared to ascorbic acid (103.03 µg mL^{-1}) that was used as a standard. Further, phosphomolybdenum method was used to assess total antioxidant capacity (TAC) that was expressed as ascorbic acid equivalents (AAE). TAC of methanolic extract of *A. wightii*, *A. racemose*, and *A. monophylla* was noted to be 655, 592, and 686 mg g^{-1} AAE, respectively.[8]

Reactive oxygen species is a term collectively used for nonradical and radical derivatives of molecular oxygen, which results in tissue injury via lipid peroxidation causing various diseases like diabetes, cancer, and atherosclerosis. Antioxidants have the potential to scavenge these free radicals and reactive oxygen species. According to a study, the protein fraction and aqueous extract from *A. monophylla* leaf possessed antioxidant properties that may result in treating different oxidative stress-related diseases. Kandappa et al.[16] determined in vitro antioxidant potential of the protein fraction and aqueous extract of *A. monophylla* leaf using various assays (DPPH, H_2O_2, ABTS, and superoxide anion assays). They reported that at 250 µg mL^{-1} concentration, the protein fraction and aqueous extract revealed 79% and 90%, 82% and 91%, 69% and 81%, and 79% and 94% scavenging activities, as determined by DPPH, H_2O_2, superoxide anion, and ABTS assays, respectively.[16]

Another species of the genus *Atalantia*, that is, *A. roxburghiana* has also been found to possess significant phenolic contents and antioxidant properties. For this purpose, a study was conducted to assess the presence of phytochemicals and antioxidant properties of methanolic extract of *A. roxburghiana* leaves. Phytochemical screening revealed the occurrence of saponins, carbohydrates, glycosides, alkaloids, phenolics, and flavonoids in the methanolic extract of *A. roxburghiana* leaves. The total phenolic contents in the methanolic extract were reported to be 39.25 mgGAE/g of extract, while the total flavonoid contents were reported to be 160.85 mg QE/g. Similarly, free radical scavenging activity of the methanolic extract was IC_{50}: 554.47 µg/mL. However, total antioxidant capacity of the methanolic leaf extract of *A. roxburghiana* was 18.73 mg AAE g^{-1}.[46] *A. monophylla* is a commonly grown small tree in the foothills of dry vegetation in different regions of Tamil Nadu. The extracts of various parts of *A. monophylla* have been used in different regions of Asia as a folk medicine. Previously, the essential oil isolated from fresh leaves of *A. monophylla* has also been explored for their chemical constituents.[8]

Later, the antioxidant potential of the essential oil isolated from leaves of *A. monophylla* has been reported using different experimentations such as DPPH free radical scavenging assay, ferrous ion chelation assay, hydroxyl radical scavenging assay, prevention of deoxyribose degradation assay, and inhibition of linoleic acid peroxidation assay. Results of this study showed that the isolated essential oil at concentration 250 µg mL^{-1} had IC_{50} values of 198.97, 204.78, 199.35, 176.54, and 219.15 µg mL^{-1} as determined by using DPPH free radical scavenging assay, ferrous ion chelation assay, hydroxyl radical scavenging assay, prevention of deoxyribose degradation assay, and inhibition of linoleic acid peroxidation assay, respectively.[47] Similarly, leaves of another species, that is, *A. racemose* have also been evaluated for their antioxidant properties. Accordingly, results of a study demonstrated that the methanolic extract of *A. racemose* leaves had IC50 values of 95.92, 40.60, and 41.81 µg mL^{-1} for DPPH radical scavenging, OH radical scavenging, and ABTS radical scavenging assays.[39] Traditionally, residents of Sri Lanka have used decocted *Atalantia ceylanica* leaves as a folk medicine for treating different diseases, specifically those related to the liver. The aqueous extract of *A. ceylanica* was lyophilized and explored for its chemical components and antioxidant properties. The reported total phenolic contents and total flavonoid contents present in the lyophilized sample were 4.87% GAE and 16.48% EGGE, respectively. The decoction of leaves of *A. ceylanica* had EC50 values as 131.2, 48.40, 163.50, and 87.70 μg mL^{-1} when determined through DPPH scavenging activity, OH-radical scavenging activity, NO-scavenging activity, and ferric ion-reducing power assay, respectively.[11] Table 11.2 shows antioxidant properties of the genus *Atalantia.*

TABLE 11.2 Antioxidant Properties of the Genus *Atalantia.*

Parameter/assay	Plant part	Results	References
DPPH assay	*A. wightii*	IC_{50}: 375.64 μg mL^{-1}	8
	A. monophylla	IC_{50}: 336.84 μg mL^{-1}	
	A. racemose	IC_{50}: 348.75 μg mL^{-1}	
Total antioxidant capacity (TAC)	*A. wightii*	655 mg g^{-1} AAE	
	A. monophylla	686 mg g^{-1}	
	A. racemose	592 mg g^{-1} AAE	
DPPH assay	*A. monophylla* aqueous extract	Percent inhibition: 79%	16
	A. monophylla protein fraction	Percent inhibition: 90%	
H_2O_2 assay	*A. monophylla* aqueous extract	Percent inhibition: 82%	
	A. monophylla protein fraction	Percent inhibition: 91%	
Superoxide anion assay	*A. monophylla* aqueous extract	Percent inhibition: 69%	
	A. monophylla protein fraction	Percent inhibition: 81%	
ABTS assay	*A. monophylla* aqueous extract	Percent inhibition: 79%	
	A. monophylla protein fraction	Percent inhibition: 94%	
DPPH assay	*A. roxburghiana* methanolic leaf extract	IC_{50}: 554.47 μg/mL	46
TAC		18.73 mg AAE g^{-1}	
DPPH assay	*A. monophylla* essential oil	198.97 μg mL^{-1}	47
Ferrous ion chelation activity		204.78 μg mL^{-1}	
Hydroxyl radical scavenging activity		199.35 μg mL^{-1}	
Prevention of deoxyribose degradation assay		176.54 μg mL^{-1}	
Inhibition of linoleic acid peroxidation assay		219.15 μg mL^{-1}	

11.4 ANTIOXIDANT PROPERTIES OF GENUS CITRUS

Genus *Citrus* is one of the most renowned genera among others under the umbrella of the family Rutaceae. Out of 392 species, only 33 are considered as accepted according to "The Plant List." This genus comprises various predominant fruits such as oranges, limes, lemons, mandarins, grapefruits, and pomelos. These fruits are predominantly liked by the natives of Asia; nevertheless, they are also abundantly cultivated in Melanesia and Australia.

Despite their unique taste, they are rich sources of phenolic compounds and antioxidants that help in promoting the health status of human beings. A study characterized free and bound phenolics and antioxidant properties in shaddock, orange, and grapefruit peels. Acetone (80%) was used for the extraction of free phenolics whereas for the extraction of bound phenolics, acid and alkaline hydrolyzed residues were treated with ethylacetate. The antioxidant activity of free phenolics was higher compared to bound phenolics in all other citrus fruit peels, except for orange peels. Maximum antioxidant activity (ABTS* assay and FRAP assay) was noticed in case of bound phenolics of orange peels, that is, 6.09 mmol/TEAC g and 71.99 mg/GAE 100 g. Whereas, shaddock peels bound phenolics possessed least FRAP (2.58 mg/GAE 100 g) and ABTS*-scavenging property (1.35 mmol/TEAC g). Furthermore, grapefruit peel-based bound phenolics revealed maximum (EC_{50}: 3.8 mg mL^{-1}) OH*-scavenging capacity; however, shaddock peel-based bound phenolics possessed minimum OH-scavenging ability, that is, EC_{50}: 16.10 mg mL^{-1}. The synergistic effect of both free and bound phenolics may contribute in developing nutraceutical products owing to their associated antioxidative properties.[29]

Citrus maxima peel extract-based silver nanoparticles have been synthesized and characterized to assess their antioxidant properties. The prepared silver nanoparticles showed significant antioxidant characteristics as determined by using DPPH and ABTS-scavenging assays.[15] Pomelo is the local name of *Citrus maxima* given by the natives of Malaysia. It has been used since ancient times as a tool in management of weight and aging by boosting the immune system responses of the body. It has also been administered to prevent constipation by promoting the process of digestion and peristalsis. Different prepared concentrations (0.03, 0.06, 0.125, 0.25, 0.5, and 1 mg mL^{-1}) of *Citrus maxima* peel in this study showed percent scavenging (DPPH assay) of 29.13, 40.29, 55.93, 61.87, 77.70, and 97.66%, respectively.[17] Table 11.3 shows the antioxidant properties of the genus *Citrus*.

Another group of workers[4] found phytochemical compounds and antioxidant properties in ethanolic extract of mesocarp, segment membrane, and pericarp of *Citrus maxima*. An initial phytochemical screening showed the occurrence of tannins, phenols, and saponins in each extract. Maximum total phenolic contents were observed in the pericarp as 25.71 mg CE/100 mL, while a minimum (13.94 mg CE/100 mL) was noticed in the mesocarp ethanolic extract. Total flavonoid contents were also in order of pericarp (31.48 mgGAE/100 mL) > segment membrane (20.17 mgGAE/100 mL) > mesocarp (11.65 mgGAE/100 mL). The antioxidant activity noticed in the mesocarp was 29.64% lipid peroxidation, while the antioxidant property of the mesocarp and segment membrane was observed as 75.95 and 71.54%, respectively.[4] Albedo is the whitish inner portion of the peel of *Citrus maxima* that is considered as a byproduct of juice processing industry. This albedo is known to have diverse polyphenolics and antioxidant properties. Nevertheless, quite few studies have been conducted on the extraction of phenols from pomelo peel through green extraction techniques. Among all the green extraction techniques, microwave-assisted extraction aids in effective heating, quick transfer of energy, cost-effectiveness, and rapidity. At power, 300 W with ethanol (60%) and mater to solvent ratio of 1:30 was the best-suited parameter for the most effective extraction of polyphenols from pomelo. At these conditions, maximum TPC and antioxidant potential was recorded to be as 2.46 gGAE L^{-1} and 1325.85 μmol TE L^{-1}, respectively.[27]

Peels of various species of the genus *Citrus* are considered functional foods owing to their antioxidant properties. For this purpose, a study was conducted to assess the antioxidant properties of fresh and frozen peels of *Citrus limon*, *Citrus aurantifolia*, and *Citrus microcarpa*. Results revealed that all three examined species had TPC and TFC ranging from 72.01 to 136.48 mgGAE/g and 50.51 to 178.32 mg QE/g, respectively. Whereas, the antioxidant properties of these citrus species (*Citrus limon*, *Citrus aurantifolia*, and *Citrus microcarpa*) were EC50: 0.82–2.70 mg/mL (DPPH assay) and 0.38–0.53 mM Fe^{2+}/g (FRAP). Result of this study showed that frozen peels of examined three citrus genera possessed more antioxidants, so can be used as a potential source of functional foods.[3] Extracts from different natural plants comprise polyphenols that possess several biological activities. Flesh and zest of *C. limon* were characterized, and their antioxidant properties were assessed using DPPH assays. At 1000 μg/mL concentration, flesh and zest of *C. limon* showed a percent inhibition of 20.3 and 71%, respectively.[26]

Different parts of *Citrus reticulata*, including juice, peel, seeds, and pulp, have been evaluated for their total phenolic contents and total flavonoids, along with associated antioxidant properties (DPPH and ABTS assays). Total flavonoid contents, total phenolic contents, and antioxidant properties varied in different parts of the citrus being examined. Reportedly, the peel of citrus had maximum amount of total flavonoids and phenolic contents as 38.97 mg RE/g and 27.18 mgGAE/g, respectively. Similarly, maximum antioxidant potential was also noticed in the peels of examined genotypes as calculated by DPPH (21.92 mg VCEAC/g) and ABTS (78.70 mg VCEAC/g) assays. Results revealed that the antioxidant properties associated with different parts of citrus were due to the significant amounts of phenolics and flavonoids present in them.[53]

TABLE 11.3 Antioxidant Properties of the Genus *Citrus*.

Parameter/assay	Plant part	Results	References
ABTS* assay	Orange peels	6.09 mmol/TEAC g	29
FRAP assay		71.99 mg/GAE 100 g	
FRAP assay	Shaddock peels	2.58 mg/GAE 100 g	
ABTS* assay		1.35 mmol/TEAC g	
OH*-scavenging assay	Grapefruit peel	EC_{50}: 3.8 mg mL^{-1}	
OH*-scavenging assay	Shaddock peel	EC_{50}: 16.10 mg mL^{-1}	
DPPH assay	*Citrus maxima* peel	Percent inhibition: 97.66%	17
DPPH assay	*Citrus maxima* mesocarp	Lipid peroxidation: 29.64%	4
	Citrus maxima segment membrane	Lipid peroxidation: 71.54%	
	Citrus maxima pericarp	Lipid peroxidation: 75.95%	
DPPH assay	*Citrus maxima* albedo	1325.85 μmol TE L^{-1}	27
DPPH assay	*Citrus limon*	EC50: 0.82–2.70 mg/mL	3
	Citrus aurantifolia		
	Citrus microcarpa		
FRAP assay	*Citrus limon*	0.38–0.53 mM Fe^{2+}/g	
	Citrus aurantifolia		
	Citrus microcarpa		
DPPH assay	Zest of *C. limon*	Percent inhibition: 71%	26
	Flesh of *C. limon*	Percent inhibition: 20.3%	
DPPH assay	*Citrus reticulata* peel	21.92 mg VCEAC/g	53
ABTS* assay	*Citrus reticulata* peel	78.70 mg VCEAC/g	

11.5 ANTIOXIDANT PROPERTIES OF GENUS CLAUSENA

Genus *Clausena* belongs to the family Rutaceae and is described as a flowering tree that is predominantly present in different areas of Africa, Asia, Australia, and the Pacific Islands. According to "The Plant List," it comprises 21 accepted species. Among these, *C. anisata* is known as a herbal plant that is most widely utilized by local practitioners in treating various diseases like lung ulceration, tuberculosis, and chronic cough in South Africa. Previously, various organic solvents have been used for the extraction of phytochemicals present in this plant. Table 11.4 shows the antioxidant properties of the genus *Clausena*.

According to a study, three different solvents, namely acetone, water, and dichloromethane, were employed for the extraction of phytochemicals from two different parts (bark and leaves) of *C. anisata*. The results of the study showed nonsignificant difference in the total flavanoid and proanthocyanidins content of bark and leaves extracts of *C. anisata*. Whereas, the total phenolic contents observed in the leaf extract of *C. anisata* were lower as compared to those present in bark extract of the same plant. The antioxidant activities determined through DPPH, nitric oxide, FRAP, and ABTS assays expressed as IC_{50} values were 0.06–0.31, 0.07–0.35, 0.18–0.26, and 0.13–0.26, respectively. Both parts of *C. anisata*, that is, bark and leaf were found to have significant antioxidative properties.[20] The reported antioxidant activities of *C. anisata* may be due to the presence of ample amounts of tannins, saponins, phenolics, proanthocyanidins, flavonoids, and alkaloids. Evidently, various degenerative problems are linked to oxidative stress conditions. The inhibitory action of *C. anisata* bark and leaf polyphenols on free radicals might validate the folkloric utilization of this plant against different infectious ailments.[20] Silver nanoparticles formulated via ethanolic extract of *C. anisata* roots have been reported to be an eco-friendly and cost-effective approach. Synthesized silver nanoparticles of roots extract of this plant showed 74.07% inhibition at a concentration of 500 μg/mL.[52]

The essential oils from *C. anisata* leaves have been traditionally used for the storage of grains in North Cameroon. The minimum concentration required for the prevention of 50% oxidation of β-carotene (EC_{50}) was documented to be 6.53 mg L^{-1} for *C. anisata* as compared to standard control, that is, BHT (524 μg L^{-1}). Whereas, the antioxidant properties of essential oils determined using DPPH and FRAP assays were 2.66 mg/L and EC_{50}: 1.77 mg/L, respectively. Depending upon the conjugated dienes profiling, essential oils possessed more antioxidant power as compared to the experimental control.[12] *Clausena excavata* is another species of the genus *Clausena* that is indigenously known as "Kemantu hitam" and is extensively used in traditional medicine in Malaysia. The methanolic extract of *Clausena excavata* leaves has been found to be a rich source of antioxidants. FRAP and DPPH assays have been used to determine the antioxidant activity of this plant. Results of a study showed that *Clausena excavata* extracted using methanol, chloroform, petroleum ether, and ethyl acetate possessed TPC and TFC ranging from 157.0–522.0 mgGAE/g and 25.80–188.60 mg QE/g, respectively. Ferric-reducing antioxidant power of *Clausena excavata* at 3 mg/mL concentration was the maximum in the ethyl acetate and methanolic extracts. Similarly, methanolic extract revealed the highest DPPH radical scavenging activities.[1] *C. excavata* is a well-renowned medicinal plant among traditional herbalists owing to its health-promoting benefits. The antioxidant properties of the methanolic extract of *C. excavata*'s leaf have been evaluated using lipid peroxidation method. Accordingly, results of a study demonstrated that the methanolic leaf extract of this plant had antioxidant activity of IC_{50}: 201.30 μg/mL. However, the water extract revealed an antioxidant activity of IC_{50}: 450.60 μg/mL. Conclusively, the antioxidant activity associated with methanol extract of *Clausena excavata* leaf was comparable to that of standard (propyl gallate), therefore validating it as a good natural source of antioxidants. Moreover, the significant antioxidant activity may be due to the occurrence of different polyphenols like gallic acid, flavanols, and angelicin in the prepared extracts of *Clausena excavata* leaf.[14]

Fruits of *Clausena indica*, another species of the genus *Clausena*, have usually been utilized as a food ingredient and folklore medicine in different tropical countries, though little work has been carried out on its phytochemical screening and bioactivities. In a study, bioguided fractions of antioxidants from two different parts, that is, seed and pericarp were extracted using ethyl acetate and hexane. For this purpose, 11 and 17 fractions were collected from hexane and ethyl acetate extracts, respectively through column chromatography to assess their free radical scavenging activity (DPPH assay) and ABTS scavenging activity (ABTS assay). Among all these, fraction 4 showed maximum scavenging activity (IC_{50}: 0.13 mg mL^{-1}), while fraction 2 exhibited the highest ABTS scavenging activity (IC_{50}: 0.31 mg mL^{-1}). Fraction T4 was found to be an abundant source of dentatin, that is, 47.32%. In a nutshell, the fruits of *Clausena indica* showed potential as a source of natural antioxidants.[19] *Clausena harmandiana* is a rich source of coumarins and carbazoles as reported in the literature. Purposely, three coumarins and nine carbazoles have been isolated from *Clausena harmandiana*. Crude extracts and isolates have been examined for their antioxidant properties through DPPH and lipid peroxidation assays. Nordentatin and 7-hydroxyheptaphylline demonstrated potential antioxidant properties on lipid peroxidation having IC_{50} values of 2.90 and 2.95 μM, respectively. However, radical scavenging activity calculated via DPPH assay revealed IC_{50} values of 29.3 and 56.8 μM, respectively. Results of these studies showed that both these isolated compounds may act as promising antioxidants and can be used as promising nutraceuticals.[43]

Ethanolic extract of *C. heptaphylla* stem bark comprises steroids, alkaloids, flavonoids, and saponins; however, it is deficient in anthraquinones and tannins. The total phenolic contents and total flavonoid contents of ethanolic stem bark extract are 13.42 mg g^{-1} and 6.90 mg g^{-1}, respectively. The ethanolic extract of *C. heptaphylla* stem bark showed significant radical scavenging capability with an IC_{50} value of 3.11 μg mL^{-1}.[10] "Wampee," scientifically known as *Clausena lansium*, is considered as a rich source of antioxidants and functional foods. Results revealed that ORAC assay values of *C. lansium* leaves were in the range of 323.1–415.8 μmol TE g^{-1} in the four developmental stages.[6] The antioxidant properties of *C. heptaphylla* peel extracts using water, ethanol, butanol, ethyl acetate, and hexane have been examined through superoxide anion, reducing power, and DPPH radical scavenging activities. Ethyl acetate extract showed the highest (95 ± 0.65%) DPPH radical scavenging activity at 50 μg mL^{-1} as compared to standard (BHT: 93 ± 1.6%). Similarly, the superoxide-scavenging potential of ethyl acetate extract at a concentration of 50 μg mL^{-1} was 88% followed by butanol fraction (47.5%), ethanol fraction (42%), water fraction (12.9%), and hexane fraction (2.5%).[31]

TABLE 11.4 Antioxidant Properties of the Genus *Clausena*.

Parameter/assay	Plant part/compound	Results	References
DPPH assay	*C. anisata* leaf extract	IC_{50}: 0.06–0.31	20
Nitric oxide assay		IC_{50}: 0.07–0.35	
FRAP assay		IC_{50}: 0.18–0.26	
ABTS assay		IC_{50}: 0.13–0.26	
DPPH assay	Ethanolic *C. anisata* roots extract	Percent inhibition: 74.07%	52
DPPH assay	*C. anisata* leaves	2.66 mg/L	12
FRAP assay		EC_{50}: 1.77 mg/L	
Lipid peroxidation assay	Methanolic *C. excavata* leaf extract	IC_{50}: 201.30 μg/mL	14
	Aqueous *C. excavata* leaf extract	IC_{50}: 450.60 μg/mL	
DPPH assay	*Clausena indica* (Fraction T4)	IC_{50}: 0.13 mg mL^{-1}	19
ABTS assay	*Clausena indica* (Fraction T2)	IC_{50}: 0.31 mg mL^{-1}	
Lipid peroxidation assay	*Clausena harmandiana* (nordentatin)	IC_{50}: 2.90 μM	43
	Clausena harmandiana (7-hydroxyheptaphylline)	IC_{50}: 2.95 μM	
DPPH assay	*Clausena harmandiana* (nordentatin)	IC_{50}: 29.3 μM	
	Clausena harmandiana (7-hydroxyheptaphylline)	IC_{50}: 56.8 μM	
DPPH assay	Ethanolic extract of *C. heptaphylla* stem bark	IC_{50}: 3.11 μg mL^{-1}	10
DPPH assay	*C. heptaphylla* peel ethyl acetate extract	Percent inhibition: 95%	31
Superoxide-scavenging assay	*C. heptaphylla* peel ethyl acetate extract	Percent inhibition: 88%	
	C. heptaphylla peel butanol fraction	Percent inhibition: 47.5%	
	C. heptaphylla peel ethanol fraction	Percent inhibition: 42%	
	C. heptaphylla peel water fraction	Percent inhibition: 12.9%	
	C. heptaphylla peel hexane fraction	Percent inhibition: 2.5%	

KEYWORDS

- **Rutaceae**
- **phytochemicals**
- **extracts**
- **antioxidant**
- **flavonoids**

REFERENCES

1. Albaayit, S. F. A.; Abba, Y.; Abdullah, R.; Abdullah, N. Evaluation of Antioxidant Activity and Acute Toxicity of Clausena Excavata Leaves Extract. *Evid. Based Complement. Alternat. Med.* **2014,** *2014.*
2. Aziz, S. S. S. A.; Sukari, M. A.; Rahmani, M.; Kitajima, M.; Aimi, N.; Ahpandi, N. J. Coumarins from Murraya paniculata (Rutaceae). *Malaysian J. Analy. Sci.* **2010,** *14* (1), 1–5.
3. Azman, N. F. I. N.; Azlan, A.; Khoo, H. E.; Razman, M. R. Antioxidant Properties of Fresh and Frozen Peels of Citrus Species. *Curr. Res. Nutr. Food Sci. J.* **2019,** *7* (2), 331–339.
4. Barrion, A. S. A.; Hurtada, W. A.; Papa, I. A.; Zulayvar, T. O.; Yee, M. G. Phytochemical Composition, Antioxidant and Antibacterial Properties of Pummelo (Citrus maxima (Burm.)) Merr. Against Escherichia Coli and Salmonella Typhimurium. *Food Nutr. Sci.* **2014,** *2014.*
5. Castello, M. C.; Phatak, A.; Chandra, N.; Sharon, M. Antimicrobial Activity of Crude Extracts from Plant Parts and Corresponding Calli of Bixa Orellana L. *Indian J. Exp. Biol.* **2002,** 1378–1381.
6. Chang, X.; Lu, Y.; Lin, Z.; Qiu, J.; Guo, X.; Pan, J.; Abbasi, A. M. Impact of Leaf Development Stages on Polyphenolics Profile and Antioxidant Activity in Clausena Lansium (Lour.) Skeels. *BioMed. Res. Int.* **2018,** *2018.*
7. Craft, B. D.; Kerrihard, A. L.; Amarowicz, R.; Pegg, R. B. Phenol-based Antioxidants and the In Vitro Methods Used for Their Assessment. *Compr. Rev. Food Sci. Food Saf.* **2012,** *11* (2), 148–173.
8. Das, A. K.; Swamy, S. Antioxidant Activity and Determination of Bioactive Compounds by GC-MS in Fruit Methanol Extracts-a Comparative Analysis of Three Atalantia Species from South India. *J. Appl. Pharma. Sci.* **2016,** *6* (2), 130–134.
9. Dhankhar, S.; Ruhil, S.; Balhara, M.; Dhankhar, S.; Chhillar, A. K.; Aegle Marmelos (Linn.) Correa: A Potential Source of Phytomedicine. *J. Med. Plants Res.* **2011,** *5,* 1497–1507.

10. Fakruddin, M.; Mannan, K. S. B.; Mazumdar, R. M.; Afroz, H. Antibacterial, Antifungal and Antioxidant Activities of the Ethanol Extract of the Stem Bark of Clausena Heptaphylla. *BMC Comple. Alt. Med.* **2012,** *12* (1), 1–9.
11. Fernando, C. D.; Soysa, P. Total Phenolic, Flavonoid Contents, In-vitro Antioxidant Activities and Hepatoprotective Effect of Aqueous Leaf Extract of Atalantia Ceylanica. *BMC Complement. Alt. Med.* **2014,** *14* (1), 1–8.
12. Goudoum, A.; Tinkeu, L. N.; Ngassoum, M. B.; Mbofung, C. M. Antioxidant Activities of Essential Oils of Clausena Anisata (Rutaceae) and Plectranthus Glandulosus (Labiateae), Plants Used Against Stored Grain Insects in North Cameroon. *Int. J. Bio. Chem. Sci.* **2009,** *3* (3), 567–577.
13. Gülçın, İ.; Oktay, M.; Kıreçcı, E.; Küfrevıoğlu, Ö. İ. Screening of Antioxidant and Antimicrobial Activities of Anise (Pimpinella Anisum L.) Seed Extracts. *Food Chem.* **2003,** *83* (3), 371–382.
14. Guntupalli, C.; Kumar, G. S.; Kumar, A. S.; Tubati, T. Evaluation of Antioxidant Activity of the Methanolic Leaf Extract of Clausena Excavata Burm. f. (Rutaceae) Using the Lipid Peroxidation Model. *Pharma. J.* **2012,** *4* (34), 22–25.
15. Huo, C.; Khoshnamvand, M.; Liu, P.; Yuan, C. G.; Cao, W. Eco-friendly Approach for Biosynthesis of Silver Nanoparticles Using Citrus Maxima Peel Extract and Their Characterization, Catalytic, Antioxidant and Antimicrobial Characteristics. *Mat. Res. Exp.* **2018,** *6* (1), 015010.
16. Kandappa, H. R.; Pillay, K.; Obulam, V. S. R.; Sharma, V. G. K.; Govender, P. In Vitro Antifungal, Antioxidant and Cytotoxic Activities of a Partially Purified Protein Fraction from Atlantia Monophylla Linn (Rutaceae) Leaf. *Trop. J. Pharma. Res.* **2015,** *14* (3), 487–493.
17. Khan, N. H.; Qian, C. J.; Perveen, N. Phytochemical Screening, Antimicrobial and Antioxidant Activity Determination of Citrus Maxima Peel. *Pharm. Pharma. Inter. J.* **2018,** *6* (4), 279–285.
18. Kulprachakarn, K.; Ounjaijean, S.; Srichairatanakool, S.; Kanjanapothi, D. Evaluation of Cytotoxicity and Antioxidant Potential of Bael Leaf (Aegle Marmelos) on Human Hepatocellular Carcinoma Cell Line. *Pharma. Res.* **2020,** *12* (3), 267–271.
19. La Anh, L.; Xuan, T. D.; Dieu Thuy, N. T.; Quan, N. V.; Trang, L. T. Antioxidant and α-amylase Inhibitory Activities and Phytocompounds of Clausena Indica Fruits. *Medicines* **2020,** *7* (3), 1–11.
20. Lawal, I. O.; Grierson, D. S.; Afolayan, A. J. Phytochemical and Antioxidant Investigations of a Clausena Anisata Hook, a South African Medicinal Plant. *African J. Trad. Com. Alt. Med.* **2015,** *12* (1), 28–37.
21. Lee, J. C.; Kim, J.; Park, J. K.; Chung, G. H.; Jang, Y. S. The Antioxidant, Rather Than Prooxidant, Activities of Quercetin on Normal Cells: Quercetin Protects Mouse Thymocytes from Glucose Oxidase-mediated Apoptosis. *Exp. Cell Res.* **2003,** *291* (2), 386–397.
22. Lewis, J. R. Biological Activity of Some Rutaceous Compounds. In *Chemistry and Chemical Taxonomy of Rutales* Academic Press: London, 1983; pp 301–318.
23. Lobo, V.; Patil, A.; Phatak, A.; Chandra, N. Free Radicals, Antioxidants and Functional Foods: Impact on Human Health. *Pharma. Rev* **2010,** *4* (8), 118–126.
24. Madhavi, D. L.; Salunkhe, D. K. Toxicological Aspects of Food Antioxidants. In *Food Antioxidants*, CRC Press, 1995; pp 281–374.

25. Lü, J. M.; Lin, P. H.; Yao, Q.; Chen, C. Chemical and Molecular Mechanisms of Antioxidants: Experimental Approaches and Model Systems. *J. Cell. Mol. Med.* **2010,** *14* (4), 840–860.
26. Makni, M.; Jemai, R.; Kriaa, W.; Chtourou, Y.; Fetoui, H. Citrus Limon from Tunisia: Phytochemical and physicochemical Properties and Biological Activities. *BioMed. Res. Int.* **2018,** *2018*.
27. Nguyen, N. H. K.; Duong, H. N.; Long, H.; Nhi, T. T. Y.; Phat, D. T. In *Effects of Microwave Extraction Conditions on Polyphenol Content and Antioxidant Activity of Pomelo Extract (Citrus Maxima (Burm.) Merr.)*, IOP Conference Series: Materials Science and Engineering (Vol. 991, No. 1, p. 012035); IOP Publishing; 2020, December.
28. Nigam, V.; Nambiar, V. S. Therapeutic Potential of Aegle Marmelos (L.) Correa Leaves as An Antioxidant and Anti-diabetic Agent: A Review. *Int. J. Pharma Sci. Res.* **2015,** *6* (3), 611–621.
29. Oboh, G.; Ademosun, A. O. Characterization of the Antioxidant Properties of Phenolic Extracts from Some Citrus Peels. *J. Food Sci. Tech.* **2012,** *49* (6), 729–736.
30. Pollio, A.; De Natale, A.; Appetiti, E.; Aliotta, G.; Touwaide, A. Continuity and Change in the Mediterranean Medical Tradition: Ruta spp. (Rutaceae) in Hippocratic Medicine and Present Practices. *J. Ethnopharma.* **2008,** *116* (3), 469–482.
31. Prasad, K. N.; Hao, J.; Yi, C.; Zhang, D.; Qiu, S.; Jiang, Y.; ... Chen, F. Antioxidant and Anticancer Activities of Wampee (Clausena Lansium (Lour.) Skeels) Peel. *J. Biomed. Biotech.* **2009,** *2009*.
32. Pynam, H.; Dharmesh, S. M. Antioxidant and Anti-inflammatory Properties of Marmelosin from Bael (Aegle Marmelos L.); Inhibition of TNF-α Mediated Inflammatory/Tumor Markers. *Biomed. Pharma.* **2018,** *106*, 98–108.
33. Rajan, S.; Gokila, M.; Jency, P.; Brindha, P.; Sujatha, R. K. Antioxidant and Phytochemical Properties of Aegle Marmelos Fruit Pulp. *Int. J. Curr. Pharm. Res.* **2011,** *3* (2), 65–70.
34. Rautela, R.; Das, G. K.; Khan, F. A.; Prasad, S.; Kumar, A.; Prasad, J. K.; ... Srivastava, S. K. Antibacterial, Anti-inflammatory and Antioxidant Effects of Aegle Marmelos and Murraya Koenigii in Dairy Cows with Endometritis. *Livestock Sci.* **2018,** *214*, 142–148.
35. Reddy, V. P.; Urooj, A. Antioxidant Properties and Stability of Aegle Marmelos Leaves Extracts. *J. Food Sci. Tech.* **2013,** *50* (1), 135–140.
36. Riso, P.; Visioli, F.; Gardana, C.; Grande, S.; Brusamolino, A.; Galvano, F.; ... Porrini, M. Effects of Blood Orange Juice Intake on Antioxidant Bioavailability and on Different Markers Related to Oxidative Stress. *J. Agri. Food Chem.* **2005,** *53* (4), 941–947.
37. Ruhil, S.; Balhara, M.; Dhankhar, S.; Chhillar, A. K. Aegle Marmelos (Linn.) Correa: A Potential Source of Phytomedicine. *J. Med. Plants Res.* **2011,** *5* (9), 1497–1507.
38. Sabu, M. C.; Kuttan, R. Antidiabetic Activity of Aegle marmelos and Its Relationship with Its Antioxidant Properties. *Indian J. Phy. Pharma.* **2004,** *48* (1), 81–88.
39. Saraswathi, K.; Mahalakshmi, B.; Rajesh, V.; Arumugam, P. In vitro Evaluation of Antioxidant and Antimicrobial Potential of leaves of Atalantia Racemosa Wight ex Hook. *Int. J. Pharma Res. Health Sci.* **2017,** *5* (6), 2031–2037.

40. Siddique, N. A.; Mujeeb, M.; Najmi, A. K.; Akram M. Evaluation of Antioxidant Activity, Quantitative Estimation of Phenols and Flavonoids in Different Parts of Aegle Marmelos. *African J. Plant Sci.* **2010,** *4* (1), 1–5.
41. Siddique, S.; Javed, S.; Nawaz, S.; Perveen, Z.; Khan, R. A.; Khanum R.; Shahzad K. Volatile Components and Antimicrobial Activity of Citrus Sinensis var. Mosammi Leaves Oil. *J. Med. Plants Res.* **2012,** *6* (11), 2184–2187.
42. Singh, U.; Kocher, A.; Boora, R. Proximate Composition, Available Carbohydrates, Dietary Fibres and Anti-Nutritional Factors in Bael (Aegle Marmelos L.) Leaf, Pulp and Seed Powder. *Int. J. Sci. Res. Pub.*, **2012,** *2* (4), 1–4.
43. Songsiang, U.; Thongthoom, T.; Zeekpudsa, P.; Kukongviriyapan, V.; Boonyarat, C.; Wangboonskul, J.; Yenjai, C. Antioxidant Activity and Cytotoxicity Against Cholangiocarcinoma of Carbazoles and Coumarins from Clausena Harmandiana. *Sci. Asia* **2012,** *38* (1), 75–81.
44. Supabphol, R.; Tangjitjareonkun, J. Chemical Constituents and Biological Activities of Zanthoxylum Limonella (Rutaceae): A Review. *Trop. J. Pharma. Res.* **2014,** *13* (12), 2119–2130.
45. Surolia, R.; Singh, A.; Bhatnagar, T. Comparative Study on the Characterization and Antioxidant Properties of Bael (Aegle Marmelos) Pulp, Shell and Seed Pectin, 2022.
46. Tania, A. Phytochemical and Biological Investigation of Methanolic Leaf Extract of Atalantia Roxburghiana, 2018.
47. Thirugnanasampandan, R.; Gunasekar, R.; Gogulramnath, M. Chemical Composition Analysis, Antioxidant and Antibacterial Activity Evaluation of Essential Oil of Atalantia Monophylla Correa. *Pharma. Res.* **2015,** *7* (Suppl 1), S52.
48. Veer, B.; Singh, R. Phytochemical Screening and Antioxidant Activities of Aegle Marmelos Leaves. *Analy. Chem. Letters* 2019, *9* (4), 478–485.
49. Wang, J.; Maldonado, M. A. The Ubiquitin-proteasome System and Its Role in Inflammatory and Autoimmune Diseases. *Cell Mol. Immunol.* **2006,** *3* (4), 255–261.
50. Wang, S. Y.; Ballington, J. R. Free Radical Scavenging Capacity and Antioxidant Enzyme Activity in Deerberry (Vaccinium Stamineum L.). *LWT-Food Sci. Tech.* **2007,** *40* (8), 1352–1361.
51. Yadav, S. S.; Dahiya, K.; Ganie, S. A.; Gulia, S. K. Antibacterial Activity of Aegle Marmelos (L) Correa. *Int. J. Pharm. Pharm. Sci.* **2015,** *7* (3), 462–464.
52. Yakoob, A. T.; Tajuddin, N. B.; Hussain, M. I. M.; Mathew, S.; Govindaraju, A.; Qadri, I. Antioxidant and Hypoglycemic Activities of Clausena Anisata (Willd.) Hook F. ex benth. Root Mediated Synthesized Silver Nanoparticles. *Pharma. J.* **2016,** *8* (6).
53. Zhang, H.; Yang, Y. F.; Zhou, Z. Q. Phenolic and Flavonoid Contents of Mandarin (Citrus Reticulata Blanco) Fruit Tissues and Their Antioxidant Capacity as Evaluated by DPPH and ABTS Methods. *J. Int. Agri.* **2018,** *17* (1), 256–263.

Index

R

S

T

W

Z